**Mathematical
Modeling
with
Computers**

Mathematical Modeling with Computers

Samuel L.S. Jacoby

Boeing Computer Services

Janusz S. Kowalik

Washington State University

Chapter 3 by
H. Blair Burner
Boeing Computer Services

Prentice-Hall, Inc., Englewood Cliffs, New Jersey   07632

Library of Congress Cataloging in Publication Data

JACOBY, SAMUEL L.S.
  Mathematical modeling with computers.

  Includes bibliographies and index.
  1. Digital computer simulation. 2. Mathematical
models. I. KOWALIK, JANUSZ S., joint author.
II. Title.
QA76.9.C65J32 1980   001.4′24   79–11870
ISBN 0–13–561555–0

To our wives,
Michal Jacoby and Krystyna Kowalik

Editorial/production supervision and interior
  design by Marianne Thomma Baltzell
Manufacturing buyer: Gordon Osbourne
Cover design by Edsal Enterprises

PRENTICE-HALL INTERNATIONAL, INC., *London*
PRENTICE-HALL OF AUSTRALIA PTY. LIMITED, *Sydney*
PRENTICE-HALL OF CANADA, LTD., *Toronto*
PRENTICE-HALL OF INDIA PRIVATE LIMITED, *New Delhi*
PRENTICE-HALL OF JAPAN, INC., *Tokyo*
PRENTICE-HALL OF SOUTHEAST ASIA PTE. LTD., *Singapore*
WHITEHALL BOOKS LIMITED, *Wellington, New Zealand*

Contents

PREFACE *ix*

Chapter 1

INTRODUCTION TO MATHEMATICAL MODELING *1*

1.1 Introduction *2*
1.2 Types, Purposes, and Usefulness of Models *3*
1.3 Types and Elements of Mathematical Models *11*
    1.3.1 Types of mathematical models according to
        their time-related behavior *12*
    1.3.2 Types of mathematical models according to their data,
        parameters, and mathematical relationships *13*
    1.3.3 Types of mathematical models according to degree
        of model refinement and structure *14*
    1.3.4 Types of mathematical models according to
        mathematical modeling problem *17*
1.4 The Mathematical Model-building and Modeling Process *21*
    1.4.1 Prototype identification and statement of the modeling problem *25*
    1.4.2 Mathematical model definition and analysis *26*
    1.4.3 Mathematical problem analysis, reformulation,
        and solution development *28*
    1.4.4 Computer program design, development, and checkout *30*
    1.4.5 Model validation, adjustment, and use *31*
References *33*

Chapter 2

THE ANALYTICAL STAGE OF MODEL BUILDING *35*

2.1   Methodology for the Analytical Stage
of Mathematical Model Building *38*
2.1.1 Prototype identification and statement of the modeling problem *38*
2.1.2 Mathematical model definition and analysis *42*
2.1.3 Mathematical modeling problem analysis and reformulation *57*
2.1.4 Solution algorithm development *63*
2.1.5 Mathematical software *71*
2.1.6 Summary *78*
2.2   A Mathematical Model of a General Two-dimensional Flow *78*
2.2.1 The prototype and the modeling purpose *79*
2.2.2 Mathematical model development *80*
2.2.3 Mathematical modeling problem analysis *83*
2.2.4 Finite-difference solution process development *85*
2.2.5 Finte-element solution process development *89*
2.2.6 Summary *93*
2.3   A Mathematical Model of Steady Nonlinear
Flow in a Pipe Network *94*
2.3.1 The prototype and the modeling purpose *94*
2.3.2 Mathematical model development *95*
2.3.3 Mathematical modeling problem analysis *99*
2.3.4 Solution process development: the implicit loop procedure *105*
2.3.5 Solution process development: the iteration procedure *111*
2.3.6 Summary *114*
2.4   A Mathematical Model for Hydroelectric
Power Systems Optimization *114*
2.4.1 The prototype and the modeling purpose *115*
2.4.2 Mathematical model development *116*
2.4.3 Mathematical modeling problem analysis: I *121*
2.4.4 Linearly constrained minimization problems *122*
2.4.5 Mathematical modeling problem analysis: II *125*
2.4.6 Solution algorithm *127*
2.4.7 Summary *130*
2.5   Review of Examples *130*
References *133*

Chapter 3

COMPUTER IMPLEMENTATION OF THE MODEL *137*

H. Blair Burner

3.1 Introduction to the Systematic Development
of Computer Programs *138*
3.1.1 The recognition that a problem exists *141*
3.1.2 The definition of the problem *142*
3.1.3 The specification of the solution requirements *142*
3.1.4 The design of the solution to the problem
that satisfies the specifications *143*
3.1.5 Adapting the solution procedure to the computer *144*
3.1.6 Construction and testing of the program as specified by the design *144*
3.1.7 Certification that the program is correct and that it satisfies the
requirements specification *145*
3.1.8 Installation of the program in the environment
of the problem solution *145*
3.1.9 Maintenance of the computer program as a correct, efficient, flexible,
economical, and useful part of the solution procedure *145*
3.1.10 Summary *146*
3.2 Computer Program Design and Adaptation
to a Computing Environment *146*
3.2.1 Preliminary remarks *147*
3.2.2 The steps in the top-down design scheme *150*
3.2.3 The subprocesses within a process *151*
3.2.4 The output sets of the subprocess *154*
3.2.5 The input sets of the subprocess *155*
3.2.6 The functional correspondence between output and input sets *155*
3.2.7 The operational sequences of the subprocesses *157*
3.2.8 Criteria for path selection *160*
3.2.9 Testing and design model *163*
3.2.10 Design evaluation *174*
3.2.11 Design documentation *179*
3.2.12 Design reviews *183*
3.2.13 Adapting the design to a computing environment *184*
3.2.14 Summary of computer program design *191*
3.2.15 Example of computer program design *192*
3.3 Computer Program Construction and Testing *199*
3.3.1 Computer program construction *199*
3.3.2 Language selection and coding; the program format *201*
3.3.3 Computer program testing *207*
3.3.4 Summary *209*

3.4   Configuration Control and Maintenance   *209*
      3.4.1 Configuration control   *209*
      3.4.2 Program maintenance   *211*
      3.4.3 Summary   *214*
References   *214*

Chapter 4

MODEL VALIDATION AND USE   *215*

4.1   Introduction to Model Validation and Use   *216*
4.2   General Consideration of Model Validity   *218*
4.3   Unreliable Computations   *220*
4.4   Analysis of Mathematical Errors in Modeling Computations   *229*
4.5   Other Reasons for Model Invalidity   *232*
4.6   Sensitivity Analysis   *233*
4.7   Model Implementation and Use   *237*
4.8   Conclusion   *240*
References   *240*

Appendix

NUMERICAL SIMULATION   *243*

A.1   Motivation for Numerical Simulation   *244*
A.2   Structure of Continuous and Discrete Simulation Models   *245*
      A.2.1 Continuous simulation models   *246*
      A.2.2 Discrete simulation models   *248*
A.3   Computer Implementation of Simulation Models and the Use of
      Special-Purpose Languages   *251*
      A.3.1 Continuous-system simulation languages   *253*
      A.3.2 Discrete-system simulation languages   *259*
      A.3.3 Evaluation and comparision of discrete simulation languages   *264*
A.4   Applications of Simulation Models   *268*
      A.4.1 System dynamics   *269*
      A.4.2 Evaluation of service systems   *273*
A.5   Summary   *279*
A.6   Sources of Information Related to Simulation Languages   *281*
References   *284*

INDEX   *287*

Preface

A *mathematical model* is an approximate representation, in mathematical terms, of a concept, an object, a system, or a process. The model behaves, in some sense, like the prototype. We use the term *prototype* here to mean anything the model represents. Mathematical models are used by scientists, engineers, operations researchers, and managers to study the behavior and operation of many different prototypes. Mathematical modeling started as an aid in scientific research and in engineering design projects. More recently, interest has focused on modeling related to operations research, natural resources, and environmental and urban development studies. Today, medical, social, and economic problems are also being studied with the help of mathematical models. The use of mathematical models has expanded in parallel with the continuing development of computers and of computational techniques.

Existing modeling experience is described in hundreds of technical papers about particular modeling projects. Several of the model-building and modeling steps are discussed in papers and books on subjects such as numerical analysis and computer programming. There is however no general book that covers *all* the steps of mathematical model building and modeling with computers, and that provides guidance and tools for those engaged in these activities. The objective of this book is to serve this need.

The book is a guide for builders and users of computer-implemented mathematical models. It is structured according to the stages of the mathematical model-building and modeling process which include: prototype identification and statement of the modeling problem; mathematical model definition and analysis; mathematical problem analysis; reformulation and solution development; computer program design and development; and model validation, adjustment, and use. To focus the subject matter, we present brief surveys of types and purposes of models and of types and elements of mathematical models. In addition, necessary and sufficient conditions for mathematical model usefulness are proposed, explored, and related to the model-building stages.

The book spans the whole process of mathematical model building and modeling, starting with the modeling objective and prototype definition and concluding with modeling and interpretation of results. Topics are discussed in the order in which the steps are normally carried out. However, the depth of coverage is not uniform because more emphasis and details are given for subjects not covered elsewhere. For the subjects not treated in great detail in this book, the reader is referred to articles and books from which he can get the information he will need. Thus this book, supplemented by the readily available reference material, is intended to be a study of and handbook or guide for mathematical model building and modeling with digital computers.

Chapter 1 includes a brief introduction to modeling, in general, and to mathematical models, in particular. Types, purposes, and conditions for usefulness of models are described and a survey is given of types and elements of mathematical models. A final section includes a description of the stages of mathematical model building and modeling, starting with prototype and modeling purpose definition and concluding with interpretation of modeling results. The iterative nature of the model-building process is emphasized. Model-building project management considerations are also discussed.

The subject of Chapter 2 is the analytical stage of model building. The chapter starts with the methodology for the identification and definition of the prototype and the modeling objective and problem; then we discuss mathematical model definition and analysis of the modeling problem, and the development of a solution technique for this problem. The discussion is illustrated with three case studies: steady, two-dimensional flow, nonlinear network flow, and a large-scale nonlinear optimization of a complex system operation.

Chapter 3 is concerned with computer implementation of the model. In it we explain the systematic design and development of computer programs and their construction, testing, and maintenance as the modeler's working tool. The nonlinear network flow model developed in Chapter 2 is used in Chapter 3 to illustrate the basic steps of the method proposed for computer program design.

Chapter 4 includes a discussion of model validation and "fine tuning," its use for modeling and interpretation of modeling results. We consider how to best use the model and how to obtain the most effective and efficient modeling strategy. The discussion then turns to the relation between modeling output and prototype behavior. This includes interpretation of the modeling results and error analysis. As in the other chapters, this discussion is illustrated with examples.

The Appendix includes a discussion on the development and use of numerical simulation models. References to related literature are given in the Appendix, as well as in each chapter.

We have worked in mathematical modeling for almost two decades and have participated in many model building and modeling projects. In this book we have drawn on our experience to present the subject to students of modeling and to future model builders and users. It is hoped that the examples and methodology presented will be helpful to workers in the many different scientific and technological disciplines in which our readers may be interested and will make a contribution to the success of their modeling projects. To facilitate this, the book is as general as we could make it, with examples drawn from many areas. The level of presentation is geared to senior undergraduate or junior graduate students, or to practitioners with some background in applied mathematics and some computer experience. Teachers who use this book as a text are referred to an article by Bonder [1973] (see references at the end of Chapter 1), which discusses educational requirements and deficiencies in the area of modeling.

The approach recommended in this text and the examples used reflect our experience with the subject. This experience could not have been gained without the help and contribution of our colleagues at Boeing Computer Services. First and foremost of these is Mr. Claude R. Gagnon, our collaborator on the projects described in Sections 2.3 and 2.4. He also read the manuscript and provided many helpful comments and suggestions. Dr. H. Blair Burner prepared Chapter 3, "Computer Implementation of the Model." We wish to acknowledge also the patience and efficiency of Mrs. Roberta A. Gillespie, Mrs. Carmel M. Neibergs and Mrs. Lucille G. Wirak in preparing the typed manuscript. Many thanks go to the management of Boeing Computer Services for supporting our mathematical modeling activities and our writing of this book.

SAMUEL L.S. JACOBY
JANUSZ S. KOWALIK

Chapter 1

Introduction
to Mathematical
Modeling

1.1 INTRODUCTION

A *model* is an imitation or an approximate representation of a prototype. The term *prototype* is used here to mean a concept, an object, a system or a process. Model behavior is in some sense similar to that of the prototype. Model uses derive from this similarity. Models are used to study, plan, design, or to control the prototype. In most cases modeling reduces cost, risk, and flow time of these tasks. Frequently, modeling is the only way to accomplish these tasks because the prototype may be unavailable or unusable for this purpose. Models are used by artists, architects, engineers, designers, planners, operations researchers, economists, managers, scientists, and many others.

In a mathematical model the representation of the prototype is symbolic, in mathematical terms including variables, parameters, and relationships such as equations and inequalities. Mathematical models frequently have to be implemented on analog, digital, or hybrid computers to make possible the data handling and calculations involved in modeling studies. Generally then, mathematical model building or adaptation and computer implementation precedes mathematical modeling studies or experiments. This book is about building and using mathematical models on computers. This process starts with the prototype: a concept, an object, a system, or a process, and a set of questions about its behavior under certain, given conditions. The objective of the process is to build a model and to use it in modeling studies to answer these questions.

A great body of mathematical modeling experience has been accumulated in the last two decades. This experience is related to the development and increased use of computers. Much of the mathematical modeling effort during this period focused on problems of the physical sciences and engineering. Mathematical models of anything from a water distribution network to the human pancreas gland have been built and studied. More recently interest has shifted to biological, medical, natural and energy resources problems, and urban development (see Gold [1977] or Brewer [1973] for example). It is also becoming increasingly apparent that mathematical modeling has the potential to make a significant contribution to the understanding and the solution of social problems (e.g., Hawkes [1973]) and to behavioral science (e.g., Lehman [1977]). It appears that the experience gained in modeling of prototypes from the physical sciences and engineering will be applicable in modeling of other prototypes in the future. Currently we are indeed witnessing a substantial diffusion of mathematical modeling methods from the so-called hard sciences (such as physics and engineering) to the socio-economic and life sciences. Eventually, socio-economic and biological processes may be accurately described by quantitative models, making possible a better understanding of these very complex phenomena.

Examples of books covering and/or illustrating certain aspects of mathematical model building and modeling include: Alexander and Bailey [1962], Andrews [1976], Bauer [1973], Bender [1978], Blackwell [1968], Dahl, et al. [1972], Emshoff and Sisson [1970], Furman [1972], Lehman [1977], Machol [1965], Smith [1968], Smith, et al. [1970], Vemuri [1978], Williams [1978], and Wolberg [1971].

The remainder of Chapter 1 introduces the reader to modeling and focuses on mathematical models. Section 1.2 is about types, purposes and conditions for usefulness of models. Section 1.3 contains a survey of types and elements of mathematical models. Section 1.4 includes a description of the process of model building, computer implementation and use. The stages of this process are justified in terms of conditions for model usefulness developed in Section 1.2.

1.2 TYPES, PURPOSES, AND USEFULNESS OF MODELS

"God created man in His image, in the image of God He created him" (Genesis 1: 27). For Rabbi Shelomo Yitzhaki (Rashi), the famous 11th-century commentator, this attribution of a human image to God was unthinkable. His interpretation of the phrase "His image" was: "In the mold which was made for man . . . made in a mold like a coin which is stamped." The creator, one might say in modern terms, used a mold to make man, a mold which had been made especially for this purpose. All other creatures were created simply by divine command, whereas the creation of man is one of the first recorded uses of modeling. Modeling has been used ever since for many creative purposes.

Models may replace a phenomenon in an unfamiliar field by one in a field with which the model user is better acquainted. With models, phenomena can be simplified, relevant properties can be extracted, and effects can be scaled up or down in space or time, to obtain an appropriate level of detail and ease of modeling experimentation. Models enable one to carry out experiments under more favorable conditions than would be possible with the prototype. In many instances these experiments cannot be carried out with the prototype. In fact modeling can, and often does, proceed even when the prototype is only a preliminary concept or design. And in general, it frequently is simpler to study its behavior through a model. The effect of changes, for example, can be studied on the model without actually implementing these changes in the prototype. Theoretical models are useful in explaining phenomena and also as a basis for physical and mathematical models. In general, modeling facilitates the achievement of many different purposes at reduced cost, risk and flow time. Current uses of modeling include observation and explanation, planning, engineering design, planning and design optimization, performance analysis, operational control, and scientific research. Different types and purposes of models are further discussed in this section. Table 1.2.1 summarizes what prototype information is available and what information results from modeling for different purposes. Table 1.2.2 indicates which model types are suitable for different purposes.

Most models fall in one of the two broad categories of physical and symbolic models. A model is said to be a *physical* (or *material*) *model* whenever the modeling representation is physical and tangible, with model elements made of materials and hardware. Examples of physical models include iconic models, hardware scale models, and analog computer models. A model is said to be a *symbolic* (or *formal*) *model* whenever the modeling

Table 1.2.1 **Prototype Information Available for and Resulting from Modeling**

PROTOTYPE INFORMATION	MODELING PURPOSE					
	OBSERVA-TION AND EXPLANA-TION	PLANNING AND DESIGN	PLANNING AND DESIGN OPTIMIZA-TION	PERFOR-MANCE ANALYSIS	OPERA-TIONAL CONTROL	RE-SEARCH
Preliminary concept, design or plan	Available	Available	Available	Available	Available	Available
Well-defined system, object or process	Resulting	Available and resulting	Available and resulting	Available	Available	Resulting
Components, configuration and fixed parameters	Resulting	Available and resulting	Available and resulting	Available	Available	Resulting
Adjustable parameters	Resulting	Available and resulting	Resulting	Available	Resulting	Resulting
Environment, circumstances, operating conditions	Available	Available	Available	Available	Available	Available
Purpose mission, appearance, behavior, performance	Available and resulting	Available	Available	Resulting	Available	Available

Table 1.2.2 Types of Models and Their Suitability for Different Modeling Purposes

MODEL TYPE		MODELING PURPOSE					
		OBSERVATION AND EXPLANATION	PLANNING AND DESIGN	PLANNING AND DESIGN OPTIMIZATION	PERFORMANCE ANALYSIS	OPERATIONAL CONTROL	RESEARCH
PHYSICAL	Iconic	×	×				
	Hardware: scale	×	×		×		×
	Hardware: analog	×	×		×	×	×
SYMBOLIC	Verbal/ logical/ drawing	×	×				
	Mathematical	×	×	×	×	×	×

representation is theoretical, or symbolic, with model elements consisting of a symbolic statement of certain structural or behavioral aspects of the prototype. Examples of symbolic models include drawings, verbalizations, logical and mathematical models and their digital computer implementations. We should remark at this point that the distinction between physical and symbolic models is not always clear. In reality symbolic models may use materials and hardware such as computers. Physical models, on the other hand, are frequently based on symbolic representations of the prototype. As an illustration of this, consider for example a set of mathematical relationships representing an engineering system which may be the basis for a physical scale model, an analog computer model (physical), or a digital computer model (symbolic).

Perhaps the simplest physical models are *iconic* models. They provide a visual representation of certain aspects of the prototype, but can be manipulated only in a limited fashion. Their use is therefore restricted mainly to static prototype studies such as architecture of buildings or sculptures. Other physical models constitute simplified representations of their prototypes, representations of those prototype properties which have been selected for study. Frequently the physical model performs the same function as the prototype in a scaled version. Such scaling can be up or down and in space, in material properties or in time.

Scale models constitute perhaps the largest class of physical models. Ship models used in towing tanks to study drag are an example of geometrically scaled models. Other fluid-flow models combine geometric and fluid property scaling. An example of time scaling in a model is the use of drosophila insects, whose reproductive cycle is short, to study heredity. Time scaling is not always possible. This is one of the major disadvantages of those physical models which perform the same, or similar, function as their prototype. Hardware scale models are frequently expensive to build and to use. They are not very flexible and therefore not suited to represent many aspects of the prototype and of its behavior. A hardware model of the fuel pipe network on an airplane, for example, cannot be subjected to all the forces and accelerations that its prototype may experience in real life. Scale models sometimes therefore represent a distorted operation of their prototype because of modeling law limitations. On the other hand, hardware scale models have many advantages including ease of communication and of validation of the modeling results, again primarily if they perform the same function as their prototype. In some cases (e.g., wind tunnel models), due to their small size and the relative expense and difficulty of other modeling approaches, the hardware scale models are the most cost-effective way to achieve the modeling purpose.

Hardware *analog* computer models constitute another large class of physical models, used for observation, planning and design, performance analysis, operational control, and research. These models are built of electric circuit elements such as resistances, amplifiers, integration networks, potentiometers, etc. Prototype attributes are represented in analog computers by analogous physical magnitudes or by electric signals. Analog models have some of the advantages of mathematical models such as flexibility, scalability, and the ability to compress modeling flow time. Hardware components used in special-purpose analog computers can be reused for other modeling purposes. Disadvantages of analog models include insufficient modeling accuracy for some applications and poor cost effectiveness. To overcome these disadvantages, digital computers have been put into increasing use as simulators of analog computers. A multitude of languages has been developed for general-purpose modeling, which makes digital computers look to the user like analog computers. This approach makes a very cost-effective tool available to the traditional users of hardware analog models. More detail about these languages is provided in the Appendix.

Examples of very simple symbolic models include geographic or topographic maps and plans of buildings and machines. These types of logical models or drawings could be termed *symbolic scale* models. The advantages and the disadvantages of these simple symbolic models are very much like those of the iconic models. They are simple and inexpensive to make, can only be manipulated in a limited way, but are very useful for the purpose of visualizing concepts, layouts and designs, or for artistic and pedagogical purposes. On the other end of the complexity spectrum we find theoretical models of *verbal* or *logical* form. These symbolic models constitute a record of theory and serve for pedagogical purposes and as a basis for other types of models.

One of the more versatile, flexible, and effective types of symbolic models is the *mathematical* model. Such models are represented in the abstract mathematical form of variables, parameters, equations, and inequalities, and do not have any physical resemblance to their prototypes. To obtain a mathematical model these key mathematical elements and relationships have to be defined and derived. The use of a mathematical model normally involves the solution of a mathematical problem consisting of these elements, relationships, and modeling data. This does often require elaborate computations and manipulation of large amounts of information. To cope with this, mathematical models are often implemented on computers. The proper choice of a computer and a solution scheme is critical to achieving the modeling purpose. The difficulties with mathematical modeling are manifold, including:

1. It may not be possible to define the mathematical relationships representing the prototype, due to inadequate theoretical understanding of the prototype.

2. Even if the relationships can be defined, they may not constitute a solvable problem.

3. Even if the problem is solvable, sufficiently detailed data for modeling may not be available.

4. Errors due to the chain of approximations and computations involved in modeling may be excessive.

5. Cost of the computations may be prohibitive.

6. It may be difficult to validate mathematical models.

These difficulties are addressed in subsequent chapters and methods to overcome them are discussed. We should note that most other forms of models have similar drawbacks and that, despite their drawbacks, mathematical models are being used at an increasing rate. In spite of the high cost of building and using mathematical models, overall cost of mathematical modeling is frequently low relative to its benefits and in comparison with other modeling methods. The cost advantage and the convenience of mathematical modeling is bound to increase with the improving cost effectiveness of computers on the one hand and the ever-increasing cost of labor and materials required for physical modeling on the other hand. Furthermore, considerable compression of flow time is usually possible with mathematical models, a favorable economic factor in many projects. Mathematical models can be used in situations which cannot be realized or simulated in any other way, for example the previously mentioned dynamic performance of an airplane fuel system in flight maneuvers prior to airplane construction. Mathematical models can be used to simulate major instabilities in systems that cannot be realized in physical models. With the advent of computer graphics, mathematical modeling can be used for observation and for design. In fact, mathematical models are ideally suited for design because of their flexibility and ease of change. Mathematical models have found an increasing number of applications in design and planning optimization. Their advantage in this case is that they enable one to compare many hundreds or thousands of options and automatically chose the best, a task no designer could do even with a regular, nonoptimizing model at his or her disposal.

The role of models in scientific research is discussed in an article by Rosenblueth and Wiener [1945] in a very general way. The importance of modeling in engineering and in planning has increased since this article was published. The article's presentation is intriguing as well as applicable to mathematical models. It is therefore rendered here with a slight change in emphasis to reflect some of the more recent applications of models. The objective of scientific research is to improve our understanding, i.e., to control or to predict the behavior of some part of the universe. Similarly, the objective of engineering projects is to improve, control, or to change some part of the universe. In management, planning and control are tools, in a general sense, for guiding and directing different processes. These management tools facilitate the achievement of objectives of, for example, scientific inquiries, engineering projects, or financial enterprises. What are some of the uses and the limitations of models in research, engineering, planning, and in related areas?

Physical models are useful because they replace a phenomenon in an unfamiliar field by one in a more familiar field. Modeling experiments are normally carried out under more favorable conditions (e.g., less time, less cost) than experiments with prototypes. In some cases experimentation with prototypes is impossible. For example, it is difficult to duplicate in a physical experiment electromagnetic impulses of a magnitude such as might be caused by nuclear detonation. In all cases physical models should be simpler and more amenable to experimentation than their prototypes.

The prediction of prototype behavior from the more favorable modeling situations presupposes a similarity between model and prototype. This similarity supposition must be based on adequate understanding and theoretic or symbolic models of both the prototype and its physical model. Rosenblueth and Wiener [1945] argue that for a physical model to be useful and productive, the symbolic model or prototype understanding on which it is based must not be weak or trivial. If it is trivial it may not suggest any nontrivial experiments for the physical model. On the other hand, the physical model may be superfluous if one cannot perform any nontrivial experiments, the results of which could not have been predicted with the symbolic model. Furthermore, if the physical model is too elaborate and less amenable to experiment than the prototype, then it is useless because it does not represent an improvement. Quite clearly the last comments apply also to the relationship between a computer implementation of a mathematical model and the basic symbolic model or prototype understanding from which it is developed. If the basis is trivial, it may not suggest any nontrivial computational modeling experiments. On the other hand, a computer implementation may be superfluous if the results which it is likely to produce could have been predicted with the basic symbolic model. If the computer implementation is too elaborate it may be expensive to use and therefore useless or not much of an improvement. It is therefore likely, according to Rosenblueth and Wiener [1945] that the following conditions are both necessary and sufficient for a physical model to be useful:

1. A phenomenon in an unfamiliar or less familiar field (prototype "space") must be replaced by a phenomenon in a (more) familiar field (model "space").

2. Modeling experiments must be carried out under more favorable conditions (including cost, flow time, etc.) than experiments with the prototype.

The same statement seems to apply also to mathematical models implemented on computers. In this case additional conditions are implied by (1) and (2) above:

1a. Computational (numerical) results must be validated and interpreted in prototype "space" (not "numerical garbage"). Otherwise the phenomenon in model "space" is not interpretable and hence unfamiliar to the model user.

2a. The mathematical modeling problem must be solvable, otherwise experiments cannot be carried out at all. We need:

Existence and stability: A solution to this problem must exist and must depend continuously on the given side conditions. Or, alternatively, any discontinuity must be understood and properly accounted for.

Uniqueness: It must be known whether or not the mathematical modeling problem admits more than one solution.

Conditioning: It must be known whether or not the mathematical modeling problem is well conditioned, i.e., whether or not small perturbations in problem data cause small perturbations in the solution.

Numerical approximation: If a numerical approximation is required (e.g., when a closed form solution is impossible or not usable) there must be a feasible computational (e.g convergent) process for such an approximation. This process must enable the model user to keep the approximation error under control.

Further implications of conditions (1) and (2) on model building and modeling will be introduced in our discussion of this process.

Rosenblueth and Wiener use the concepts of "closed" and "open" boxes to illustrate the application of theoretical models. In research and engineering one frequently encounters problems in which a fixed finite number of input variables determines a fixed finite number of output variables. The problem is determinate when the relations between these finite sets of variables are known. In general, however, these relations are not necessarily unique: the same output may be obtained from the same input with different relations, that is, with different physical prototype mechanisms. Suppose that the set of relations, or the physical "system mechanism" were enclosed in a box that could be approached only through input and output terminals. This box would be indistinguishable from other boxes capable of producing the same outputs from the same inputs. Distinction between these "closed" boxes may be made possible only by addition of input and/or output terminals. Thus the relation between the input and the output may become better known, or—one might say—the box may become more "open." The more terminals available, the more open the box. Therefore the only difference between an open and a closed box is one of degree, i.e., a quantitative and not qualitative difference. Many scientific investigations, and engineering designs, plans, etc. begin as closed box problems; only a small part of the significant variables and relations are recognized initially. As the project progresses two types of developments occur. New inputs and outputs are added and the closed boxes are thereby gradually opened. At the same time, however, the number of closed boxes may actually increase by further subdivision of the original boxes that are opened. This is a top-down process in terms of the previously described hierarchy of scientific questions or engineering designs. In this process we are moving from the relatively simple, more abstract and general model, to the more complex and concrete theoretical model. At any intermediate stage of the process the scientist or engineer has a collection of model elements. For some of these he knows the detailed relation between input and output variables, while for others only the functional performance or behavior is known.

This theoretical model of the modeling process may be useful in explaining and recording observed facts and may serve as:

1. A basis for building and using physical and mathematical models,
2. A basis for design synthesis,
3. A pedagogical device.

Thus, according to Rosenblueth and Wiener [1945], progress on scientific (and we add: also other) projects can be seen as sequences of theoretical-symbolic models and their physical-mathematical implementations. In scientific research this leads eventually to formal theory whereas in engineering design it leads to the physical prototype. Is the modeling process ever carried to the limit? Physical models start as rough approximations of the real-life situation under study, and gradually approach the complexity of the prototype. The model will tend to become identical with the prototype, and as a limit will become the prototype itself. This limit is obtained in certain engineering design situations. In research, however, it may be impossible to go to the limit. This is, for example, the case in medical research where animals and not humans must be used as models. But is it necessary to go to the limit? Consider the question in light of the stated conditions for usefulness of a physical model. Should a physical model fully realize its purpose, the prototype could be understood in its entirety and a physical model would be superfluous. According to Rosenblueth and Wiener the situation is the same with theoretical models. The ideal theoretical model, the limit of the modeling process, would cover the entire universe. It would be of the same complexity and have a complete one-to-one correspondence with it. Any scientist capable of developing and comprehending such a model would find the model unnecessary because he could grasp the universe as a whole. This limit probably cannot be achieved, a fact which does not, however, detract much from the usefulness of models.

1.3 TYPES AND ELEMENTS OF MATHEMATICAL MODELS

In this section we introduce the elements and types of mathematical models. We also discuss noteworthy characteristics of elements of models of certain types. Mathematical models in general consist of a model interior, a boundary, and a set of boundary and initial conditions. These conditions are imposed on the model and represent aspects of the prototype environment which are relevant to the modeling experiments. The model interior is often structured, consisting of interconnected components, each of which represents a component of the prototype. In general, models have the following elements:

1. *Variables*, which may assume any value in a given set of values; we distinguish between independent and dependent variables.

2. *Fixed parameters*, are variables which may assume only one, fixed value in any particular modeling experiment. Sometimes variables are fixed in a modeling experiment and are therefore, by this definition, parameters.

3. *Mathematical expressions* such as equations or inequalities combining model variables and parameters.

4. *Logical statements* about the model.

5. *Data* such as tables of numbers or graphical information.

Mathematical models can be classified according to several different criteria, including:

1. *Time-related behavior* of the model;
2. Type of model *data, parameters* and mathematical *expressions*;
3. Degree of model *refinement* and model *structure*;
4. *Mathematical* modeling *problem* type.

In the remainder of this section we analyze the different types of models in terms of these classifications.

1.3.1   Types of Mathematical Models According to Their Time-related Behavior

Mathematical models are said to be *dynamic* or *unsteady* if their behavior varies with time, and time therefore enters as an independent variable. Models are *static* or *steady-state* if their behavior is constant and does not vary with time. Models are *quasi-steady-state* when an unsteady prototype behavior is represented by a sequence of steady-state modeling experiments. Two types of unsteady models exist: The first are *instantaneous* models whose behavior at any given instant in time depends only on circumstances or factors present at the same instant. The second and more general type of unsteady models are *memory* models whose operation at any given instant may depend on circumstances or factors present at times prior to this instant. The memory models are said to be *time-invariant* (*stationary*) if their memory extends over a time interval of a given finite length prior to the instant of operation, independent of the instant of operation itself. Memory models are not time invariant if their memory extends to a fixed time instant in the past, or to $-\infty$ at any given time. This leads to the question of causality: Models are *causal* if their operation at any instant depends only on the past and present, and *noncausal* if it depends also on the future. Although physical systems are generally causal, the behavior of certain engineering systems may be considered in a more general light which leads to noncausal mathematical models. More on this subject can be found in an article by De Santis and Porter [1973].

Unsteady models may be used for modeling steady as well as unsteady behavior of prototypes. The reason for using an unsteady model for a steady prototype could be computational. A common solution strategy for the steady-state modeling problem, for example, is by means of an unsteady problem. The unsteady problem is solved with this method and the solution is allowed to approach the desired steady state. Unsteady or quasi-steady-state models are said to be *recursive* if their state (i.e. behavior, performance or operation) at any given instant or time interval depends on their state at an earlier time instant or time interval (compare this to the definition of multistage models given below). A recursive model is a memory model, discretized with respect to time. Modeling with recursive models usually proceeds with time because the state of the model at any given time is not defined unless its state at the earlier time has been determined. It is possible, however, to devise modeling solutions treating all stages simultaneously. This is illustrated in the example of a mathematical model for hydroelectric power optimization, one of the case studies discussed in Chapter 2.

1.3.2 Types of Mathematical Models According to Their Data, Parameters, and Mathematical Relationships

Elements of mathematical models include variables, parameters, mathematical relationships such as equations and inequalities, logical statements, and data. Different types of mathematical models can be distinguished depending on the nature of these elements.

Mathematical models are said to be *deterministic* if their elements are sufficiently specified so that the model behavior, performance, or operation is exactly determined. Mathematical models are *stochastic* if this is not the case, and uncertainties or stochastic data or elements are involved, or parameter values are known in terms of probability distributions. In this case model behavior, performance, or operation is probabilistic.

Mathematical models are *continuous* if all their data, parameters, and relationships are continuous and are *discontinuous* or *discrete* otherwise. Models can be continuous at all times or only at certain time instants. Models are discontinuous if interactions between their variables occur at discrete times only, separated by intervals of no interaction. Models involving stochastic effects, for example, are discontinuous.

Mathematical models of processes may be classified according to whether the process is a *flow* or a *batch* process. Flow, or continuous processes give rise to continuous models. These processes operate in a continuous fashion on a continuous input and produce a continuous output. Batch, or discontinuous processes give rise to discontinuous models. These processes may be discontinuous or involve discontinuous (intermittent) inputs and/or outputs. Discontinuous processes are frequently periodic or cyclic and may be modeled as such.

Mathematical models of processes may also be classified according to their parameters. The parameters in a *fixed-parameter* model are determined once and for all and are not modified throughout the modeling process. The parameters in an *adaptive-parameter* model, on the other hand, may be changed and thereby adapt the model behavior to changes in prototype behavior. Such changes may occur, for example, in a process as its environment or as properties of materials involved in it change with time.

1.3.3 Types of Mathematical Models According to Degree of Model Refinement and Structure

We distinguish between distributed and lumped models, (Smith et al. [1970]). In a *distributed* model one or more independent variables denoting spatial position (or degree of freedom) are involved, and the dependent variables and other model relationships depend on spatial position. Model relationships are based on changes of dependent variables corresponding to infinitesimal changes in independent variables. *Lumped* models are "zero-dimensional," their variables and modeling relationships do not depend on spatial position. Instead, average, or otherwise representative, values of variables are used in model components. Model relationships are based on differences between these representative values. The distributed model is suitable for so-called continuum or field problems. It is obviously more detailed than the lumped model. Distributed models normally lead to modeling relationships in the form of differential equations. Lumped models lead to difference equations. It should be pointed out, however, that differential equation models are often approximated and replaced by difference equation models in their computer implementation. Therefore, the final tool used for modeling experiments may not be more detailed than if a lumped model had been defined at the outset. In fact, as we show in one of the examples discussed in Chapter 2, exactly the same computer implementation may result from a distributed and a lumped-model definition.

A model is said to be *complex* if it is a combination of interconnected and interacting "sub"-models, or components. Often complex models have distributed as well as lumped-component submodels. Complex models can be steady- or unsteady-state, the former being a special case of the latter, more general case. Unsteady complex models are used in control systems design. Steady complex models are used in economic analyses for example.

Prototypes with a special structure are usually modeled as structured complex models. It is very important to recognize which model structure may be applicable, because the structure often gives rise to modeling simplifications. Model structures that can be taken advantage of are tree and network structures. The model is said to have a *tree* (or *branching*) *structure* if its components are interconnected in a manner such that no closed loops are formed (e.g., a system of rivers and tributaries). A model is said to have a *network* (or *cyclic*) *structure* if its components are interconnected in such a way that closed loops are formed. The interconnections or the interfaces between the components which constitute the model structure are represented by an incidence matrix or a set of linear algebraic equations with coefficients of magnitude zero or one (a linear graph). These equations represent the (Kirchhoff) network laws stating continuity and single-valuedness of variables at the points of interconnection (nodes). We consider these laws in greater detail below.

Two types of independent variables can be distinguished in distributed models: the *spatial* or *position variable* and *time*. Accordingly these models can have one, two, or three spatial dimensions and a fourth, or time "dimension." Lumped models do not have spatial variables but time may appear as an independent variable in these models. We also distinguish two types of dependent variables: across and through variables (for details see Section 2.1.2):

1. *Across variables* are scalars relating the condition at one point to that at some other point. An across variable is usually measured as a (potential) difference between values at two points in the model or between the value at one point in the model and at one reference point such as ground.

2. *Through variables* are vectors representing the flux of some quantity traversing an elemental cross section of the model.

Three basic types of elements (or "element characteristic functions") appear in distributed and lumped models, (Smith et al. [1970]). They represent different phenomena in terms of relationships between across and through variables, other dependent variables, and fixed parameters. These elements also have an orientation. In distributed models the orientation is that of a (set of) spatial coordinate(s). Lumped-model elements have an initial (input) node and a terminal (output) node. The three types of elements are (for details see Section 2.1.2):

1. Dissipators,
2. Reservoirs of flux,
3. Reservoirs of potential.

Dissipators conduct a flux of a magnitude that is proportional to a potential gradient, or difference across the element. The factor in the proportion is referred to as *dissipativity* (or its inverse—*conductivity*) per unit of length.

In *reservoirs of flux* (or of energy) the rate of change of flux with respect to time is proportional to the potential gradient or difference across the element. The factor in the proportion measures the capacity of the flux reservoir per unit length in the flow direction.

In *reservoirs of potential*, the rate of change of flux with respect to distance is proportional to rate of change of potential with respect to time. The factor in the proportion measures the capacity of the potential reservoir per unit length in the flow direction.

Complex models consist of interconnected component models. The component models, which may be distributed or lumped, have initial and terminal nodes. This implies a direction that is associated with each component model. Two or more component models are interconnected if they have a common node. A *path* is said to exist between two nodes if it is possible to trace a line from one to the other without losing contact with interconnected components. A *circuit* (or a *loop* or *mesh*) is a set of interconnected components which form one and only one closed path, i.e., a path whose initial and terminal nodes are one and the same. The relationships between through and across variables in complex models are referred to as the two *Kirchhoff laws* of network flow (more detail in Section 2.1.2):

1. The *first Kirchhoff law* is the *node law*: it states that the oriented sum of the through variables at every node is zero.

2. The *second Kirchhoff law* is the *circuit law* or *loop law*: it states that the oriented sum of the across variables around every circuit of interconnected components is zero.

Another type of structured model is the feedback/feedforward or multistage process model. Processes usually have inputs and outputs, and so have their models. Mathematical models of a process are said to be *multistage* if the input of any stage of their operation depends on the output of the previous stage. (Compare this with the definition of recursive models given above.) The multistage models may have a *series* or a *branching* (tree), or a *cycling* (network) structure.

Hyvärinen [1970] distinguishes the following types of variables in process models:

1. *Controllable variables* are independent variables whose values can be measured and controlled.

2. *Uncontrollable variables* are independent variables whose values can be measured but not controlled.

3. *Unknown variables* are independent variables whose values can neither be measured nor controlled.

4. *Performance variables* are dependent variables which define process outcome.

5. *Intermediate variables* are performance variables of subprocesses (and therefore also dependent variables).

A mathematical model of a process is said to be a *feedback model* if its input depends on its output. A model is said to be a *feedforward* (or *straight-through*) model if its output depends only on its input and no feedback exists. The purpose of the feedback is usually to control the model performance (i.e., output). The feedback control is said to be *automatic* if it is accomplished without human intervention. To achieve the control, the value of the controlled variable (output) is compared to a *reference value* or *condition* (or *ideal* or *desired condition*). The difference between the two variables is used to control the "controlled" variable. In certain engineering disciplines it is customary to use lumped "input–process–output" (or "transfer function") models. The basic elements of these models include:

1. *Control elements*, which have as input an actuating variable that depends on the reference condition and the feedback output. The control element output is input into the

2. *Controlled elements*, the output of which is the controlled variable. It goes into the

3. *Feedback elements*, the output of which is fed into the control element as previously indicated.

The purpose of this design is to control the controlled variable, as indicated.

The motivation for the use of this modeling approach in general is that it facilitates the application of results from (linear) control theory. Thus the behavior of systems modeled in this manner is easier to understand by those who are familiar with linear control theory. Terminology and other details about this type of model can be found in Shinners [1964].

1.3.4 Types of Mathematical Models According to Mathematical Modeling Problem

As defined above, mathematical models constitute approximate representations of certain elements, properties, and attributes of a prototype. Models behave, in some sense, similarly to the prototype, and are used to study prototype behavior. In order to be useful, models must lead to solvable mathematical problems. The study of prototype behavior is accomplished with the help of modeling experiments in which these mathematical problems are solved. The type of mathematical problem solved in the modeling experiments depends primarily on the prototype and the modeling purpose. But the original problem may be reformulated once or more in the process of defining and building the mathematical model. Obviously the mathematical modeling problem may change each time the model is reformulated. This is illustrated in the examples discussed in Chapter 2: In one case a distributed model is formulated at the outset leading to a boundary value problem involving a partial differential equation. The next step replaces this model by a difference (algebraic) equation approximation, a solution to which can be computed by successive approximations. It is interesting to note that a lumped model of the same prototype leads to precisely the same computer implementation. In another example, discussed in Chapter 2, we formulate a multistage optimization problem. This could, at least theoretically, be cast as a dynamic programming problem. Instead we reformulate it as a nonlinear programming problem, because in this form it admits a solution algorithm that is more economical in computer resources.

One very important type of model is the *simulation model.* We devote a special appendix in this book to simulation models. In one type of these models the state of certain system or process prototype features, activities, or events is represented and kept track of. The model representation is usually a computer program. These models may be used when:

1. The prototype mechanism is not sufficiently understood to derive detailed mathematical modeling relationships representing it.

2. Modeling relationships that could be derived would be intractable.

3. Data needed for modeling experiments with another kind of mathematical model are lacking.

4. The modeling purpose can be satisfied with a logical simulation model.

Sometimes the mathematical model formulation representing the prototype mechanism in detail leads to a problem that can be solved *analytically,* in "closed form." In many cases, however, the closed form solution is not usable for modeling experiments. It may, for example, consist of a series, whose terms must be computed. If many terms are required for sufficient solution accuracy, it may be more economical to reformulate the problem as a numerical problem. Many mathematical models representing the prototype mechanism in detail also lead to *numerical* problems. To solve most of these problems, a sequence of successive approximations to the solution must be computed iteratively. Each approximation is "better" than its predecessor. The type of numerical modeling problem involved in the modeling experiments depends on the original model formulation and on its reformulations. The problems could be linear or nonlinear of the following main types:

1. Equations: algebraic, transcendental, differential, integral;

2. Eigenvalue problems;

3. Approximation such as interpolation, numerical integration, and differentiation;

4. Optimization;

5. Methods involving artificial data such as Monte Carlo methods.

It is useful to consider briefly the types of mathematical relationships, and hence problems, arising in different model types. We have already observed that distributed models usually involve infinitesimal variations in a continuum. This leads to *differential equations*. Depending on the number of spatial variables (degrees of freedom) in the model, these could be ordinary or partial differential equations. In general, unsteady models, distributed or lumped, involve derivatives with respect to time and also lead to differential equations. Thus ordinary differential equations occur in:

1. Distributed steady-state models, if the dependent variables are functions of position in one dimension,

2. Distributed unsteady-state models, if the dependent variables are constant through the model, or

3. Lumped unsteady models.

Partial differential equations result in:

1. Distributed unsteady models with one or more independent variables, or

2. Distributed steady models, with two or more independent variables.

It can be shown that there is a connection between the types of elements in the model and the type of resulting differential equation (more on this in Chapter 2). To be solvable, models involving differential equations must also include side conditions: initial conditions for unsteady problems with derivatives with respect to time, and boundary conditions for problems with derivatives with respect to spatial coordinates. *Side conditions* specify either the dependent variable or a derivative of it for some given value of one independent variable, and every value of the other independent variables. Depending on the case, then, the side conditions are either differential or algebraic equations. The number of conditions needed for the problem to be solvable, depends on the order of the derivatives and on the number of independent variables appearing in the differential equations; the required number of conditions, for each dependent variable in terms of one of the independent variables, equals the order of the highest derivative with respect to the given independent variable.

Differential equation models are usually reformulated and implemented on the computer in terms of difference equation approximations. Thus the problem actually solved in the modeling experiments is a set of *algebraic equations*. In complex and in lumped models, algebraic equations (linear) represent the interconnections between the component models. As we have seen, for example, the Kirchhoff laws of network flow lead to linear algebraic equations representing continuity in through variables and single-valuedness of across variables at the interconnections. Component models, relating across and through variables in the model elements, can be algebraic, trans-cendental, or differential equations (the latter, for example, in unsteady situations).

Nonoptimizing models are used to study the behavior of prototypes which are completely determined. In this case the model represents the prototype, and the "input" of the modeling experiment defines the conditions under which prototype behavior is being studied. These conditions are represented in terms of parameter values, boundary, and initial conditions. The "output" of the modeling experiments defines model behavior. It includes the values of the model variables which solve the modeling problem. *Optimizing* models are used to study the behavior of prototypes that are not completely deter-mined initially. The modeling experiment defines model behavior and also completes prototype determination. Optimizing models are used for example to determine optimal designs. That is, certain design features are determined in the modeling experiment, as is the behavior of the model corresponding to each design. Optimizing models must have one element that nonoptimizing models do not have. This element is an *optimality criterion* (*criterion* of *merit* or *objective function*). It is a mathematical function relating the model vari-ables and parameters, whose value measures a degree of optimality or merit, corresponding to each set of values of these variables. Another element present in many optimizing models (and normally absent in nonoptimizing models) are inequality relationships involving the model variables and param-eters. The modeling problem that includes inequalities frequently admits more than one, so-called, "feasible solution." These are sets of values of the independent variables satisfying all model equations and inequalities. The modeling exercise consists of finding the optimal feasible solution, i.e., that which also satisfies the optimality criterion. The "input" of an optimizing modeling experiment is similar to that of a nonoptimizing modeling experi-ment. It defines the conditions under which the prototype behavior is being studied. The "output" is again the model behavior, but in addition it defines the optimal prototype represented by the model. Optimal is understood in the sense defined by the optimality criterion.

Optimizing models lead to optimization problems. If the modeling rela-tionships and the objective function are linear, the optimization problem is linear. This type of problem is usually referred to as a linear programming problem. If either the modeling relationships and/or the objective function are nonlinear the optimization problem is nonlinear, or a nonlinear program-ming problem. Optimal programming, or functional optimization problems, arise if the modeling relationships involve differential equations and the optimality criterion is a functional. More about this subject can be found in Jacoby et al. [1972]. Optimization models of interrelated multistage processes can be formulated as so-called dynamic programming problems. Frequently large-scale optimal programming problems and dynamic programming problems are reformulated as nonlinear programming problems and solved as such, because currently available techniques can better handle the latter type of problem. Frequently very large-scale nonlinear or linear programming problems are decomposed into a set of interconnected (and in some sense) simpler problems.

1.4 THE MATHEMATICAL MODEL-BUILDING AND MODELING PROCESS

The process of mathematical model building and modeling is an iterative process. The modeler often cycles through it more than once. He gets closer to achieving the modeling purpose with each cycle while, at the same time, refining the definition of the modeling purpose. The stages of this creative and involved process are frequently pursued simultaneously, or at least all other stages are being considered during any one stage. The process can be subdivided into the following steps (compare with Kwatny and Mablekos [1975]):

1. Prototype identification and statement of the modeling problem.

2. Mathematical model definition and analysis.

3. Mathematical problem analysis, reformulation and solution development.

4. Computer program design, development, and checkout.

5. Model validation, adjustment, and use.

The components of the model-building and modeling process are illustrated in Fig. 1.4.1, and the iterative and feedback/feedforward nature of the process is illustrated in Fig. 1.4.2. In this section we provide a preliminary and general discussion of the steps of the process, and highlight their most important aspects. The discussion uses the previously stated general necessary and sufficient conditions for model usefulness (see Section 1.2) and develops equivalent conditions applicable to the individual steps. These equivalent conditions motivate and provide guidance to the definition of the tasks that constitute these steps. More detailed discussions of the above steps (1)–(3), step (4), and step (5) are presented in Chapters 2, 3, and 4 respectively. A guide to where these discussions can be found is presented in Fig. 1.4.3.

Before proceeding with the subject of the section we would like to turn our attention briefly to the management aspects of mathematical model-building and modeling. Whether the project is large and complex, or simple,

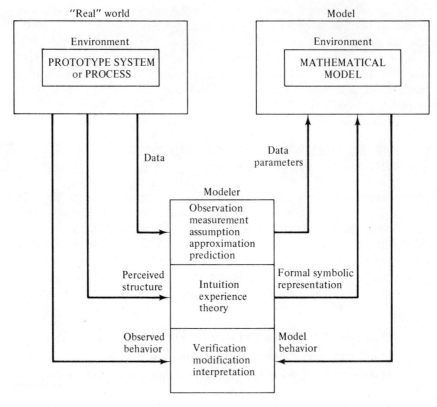

Figure 1.4.1 Components of the modeling process.

involving as little as one worker, success is more likely with a project management plan and project control. The elements of a project management plan normally include:

1. A technical plan including task breakdown and definition.

2. Realistic schedules and milestones.

3. A resource utilization plan (resources include technical workers, computer facilities, etc.).

4. A budget plan.

5. An overall management, review, coordination, and control plan.

The plan should be established at the outset and reexamined periodically. This will insure that the project is discontinued if its purpose is unachievable, that the project scope is modified if necessary, or that the purpose is achieved as planned. Constraints on resources, budget, and project milestones must be satisfied to insure project success.

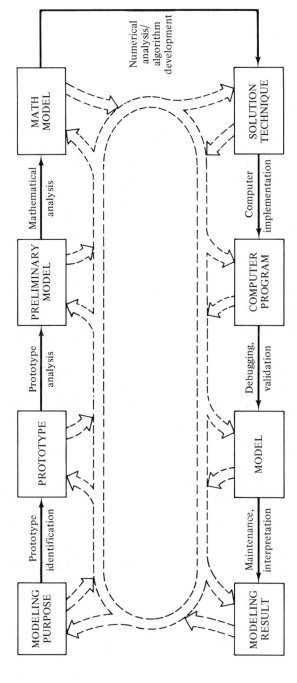

Figure 1.4.2 Iterative nature of model building and modeling in practice.

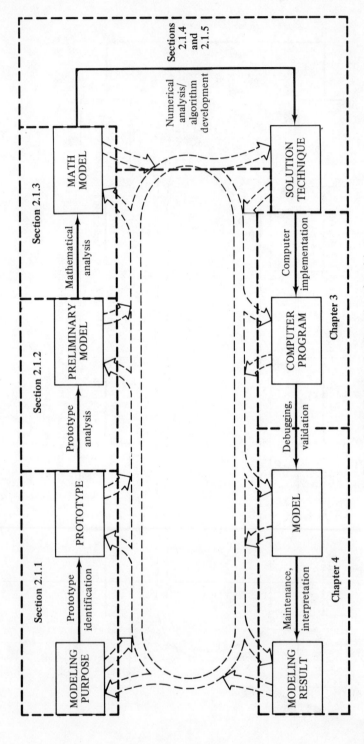

Figure 1.4.3 Guide to the discussion of model building and modeling in Chapters 2–4.

1.4.1 Prototype Identification and Statement of the Modeling Problem

Before beginning to develop a model we have to recognize the modeling problem and the need for its solution. The modeling problem usually arises within a bigger problem or project, such as a scientific investigation or an engineering project. The scope of the model must be understood in this context and the solution must interface with the other project elements. We should establish the modeling purpose, availability of data, a preliminary modeling strategy, and the performance measures of the model within this context. If this is not done, the modeling problem may be inadequately motivated or its solution may be useless, rendering the model useless in either case.

In large-scale projects the initial problem statement is fairly extensive. It addresses the following issues:

1. Purposes of the problem investigation and questions to be answered by a model (model motivation).

2. Scope and magnitude of the problem to be solved.

3. Data and other information available.

4. Anticipated depth of investigation.

5. Limitations and boundaries of the problem definition, as well as potential extensions.

6. Breakdown of the problem into smaller tasks.

7. Management plan including: schedules, milestones, personnel, computer requirements, etc.

8. Model performance criteria.

At this stage of the modeling, the applicable solution methodology is normally not known, but we should have in mind candidate techniques capable of solving the problem. In fact the awareness of the existing methodology for problems in the field provides useful guidance to problem formulation and scoping. The initial problem statement will seldom remain unchanged during the project. We expect that it will become better defined and understood as work on the project progresses.

To summarize, in this initial stage we obtain a tentative definition of the prototype and the modeling purpose. We also compile a list of the information available to support the achievement of this purpose, and consider candidate solution strategies. This is accomplished within the context of the overall problem or project which the modeling solutions will support.

1.4.2 Mathematical Model Definition and Analysis

In this step we define the model elements and their interrelationships and analyze the resulting combination. A mathematical model is a set of quantitative and logical statements representing the relevant features of the prototype of concern. The model never can or should replicate reality exactly; only features relevant to the modeling purpose contribute to model definition. A model is always an approximation of limited accuracy and validity. All model uses, however, are predicated on the similarity between the prototype and the model.

There are at least five reasons for the approximations inherent in every model:

1. Need to simplify and reduce the degree of complexity of the real phenomena under investigation.

2. Inaccurate measurements and data.

3. Imperfect hypotheses which reflect our frequently imperfect understanding of the prototype's internal mechanism.

4. Modeling approximations.

5. Scaling, up or down, in space or time.

In the process of selecting the model type we are guided mainly by the objectives of the modeling study, the prototype characteristics, limitations of existing solution methodologies, and availability of data. Frequently we can choose among several different approaches with distinct merits and disadvantages.

The common elements of mathematical models which need to be defined at this stage are:

1. Variables and their ranges of variation.

2. Fixed parameters.

3. Mathematical expressions, e.g., equations, inequalities, constraints, optimization objectives, etc.

4. Logical statements defining properties of the model or solution.

5. Boundaries and boundary and initial conditions.

6. Data, model input and output.

7. Other quantitative components of the formulation: geometrical assumptions, graphs, diagrams, etc.

Variables are the model elements that can assume more than one value during the solution process. The solution we usually seek in modeling experiments consists of the values of the variables satisfying the modeling expressions. The choice of variable types depends on the type of model we deal with; i.e., variables may be discrete, continuous, random, dynamic, etc. Fixed parameters, which do not change during the process of any one modeling experiment, may nevertheless be changed from one experiment to another. Every variable or parameter has its symbol, units of measurement, and range of values.

Elements (3)–(7) above represent formally the perceived structure of the observable prototype under consideration. To express our understanding of the model structure and function we use available theory and experience, as well as intuition. In the development of models of "physical" systems, for example, we apply laws of conservation and behavior in control volumes. The usefulness of the model is predicated on proper theoretical understanding of the prototype, and on sufficient data. If the prototype is not well understood, or insufficient data is available, no nontrivial modeling experiments are possible. If the prototype behavior is completely understood the model may be superfluous. It is also important to attain a proper degree of model refinement: excessive model realism and detail may lead to excessive computation and possibly to error-prone and uninterpretable modeling results. Excessive simplicity and aggregation may make no meaningful modeling experiments possible. We remind the reader that these considerations are tied to the general, necessary, and sufficient conditions for usefulness of computer models (see Section 1.2).

In general we can distinguish between real and artificial modeling data. Real data are obtained by actual measurements and statistical observations. Artificial data are generated in lieu of real data, typically by using random number generators, interpolation and extrapolation techniques or simply by making reasonable guesses. Availability of data and their accuracy not only impacts the quality of results obtained from modeling, but greatly contributes to the choice of model. The selection of the mathematical formulation, as well as degree of simplification is directly related to the information available for modeling definition and exercise. This relationship between data and model structure is in the very heart of the model construction process.

To summarize, in this stage we obtain a definition of the model type and elements, including variables, parameters, mathematical relationships, boundary and initial conditions, any other representation of the prototype behavior and data to be used in modeling. At this point we usually return to the previous stages of the model-building process to examine if there is any reason to believe that our modeling purpose is not achievable with the model as defined. If that is the case, modeling purpose as well as model definition may have to be adjusted. If the purpose appears to be achievable we can proceed to analyze and solve the mathematical modeling problem. This is the problem of finding the values of the model variables satisfying all the modeling relationships, boundary and initial conditions.

1.4.3 Mathematical Problem Analysis, Reformulation, and Solution Development

Several questions must be explored before proceeding to develop a solution method for the mathematical modeling problem, including:

1. Is the stated problem solvable and, if it is, does it have a unique solution?

2. Does the problem solution depend continuously on boundary and on initial data, or alternatively, have discontinuities been understood and properly accounted for?

3. Is the problem well conditioned?

4. Can the problem be replaced by an equivalent or approximately equivalent but simpler problem reformulation?

Unsatisfactory answers to these questions may imply that our modeling project is headed for failure; if the modeling problem is not solvable, stable (condition (2) above), or replaceable by a simpler reformulation, modeling experiments are impossible.

At first we resolve the question of existence of a solution, in general, to the stated problem. If the solution is infinite, or otherwise mathematically undetermined, the problem does have to be reformulated. If the solution exists, it may be unique or not. If it is not unique, the consequences of multiple, possibly infinitely many, solutions must be evaluated. Frequently, but not always, the modeling purpose can still be achieved under these circumstances. Next, problem stability and conditioning must be investigated. Here we note that boundary and initial data are often inaccurate. This may cause very large solution errors unless the problem is well conditioned. The next question is whether or not the problem is replaced by an approximation in the solution process. If it is, the equivalence or the degree of closeness of the approximation solutions and the approximated problem must be established. Also, we would like to retain as much as possible of the original problem structure in order not to cause modeling results to be uninterpretable (e.g., by oversimplification).

The development of mathematical solution technique (algorithm building) is related to problem definition and the available data. This development is either parallel to the model formulation stage or at least overlaps with it. Taxonomy of mathematical solution techniques would be very extensive, but in general we can distinguish three types of solution techniques: analytical, numerical, or mixed. The scope of problems that can be handled analytically is rather limited. It can be broadened considerably by the use of numerical techniques. In many cases analytical solutions to modeling problems have to be evaluated numerically to achieve the modeling purpose.

Computers are often employed in this process. Numerical and mixed techniques are of more interest to us since they are suitable for the majority of the modeling problems and lend themselves to computer implementation. Special numerical experimentation procedures referred to as simulation methods (see Section 1.3.4) bypass the mathematical representation of the prototype mechanism. Instead, the simulation model tracks the prototype behavior (state, activities, and/or events) and reports it to the model user, either at predetermined time intervals or following the occurrence of certain events. Utilization of this concept combined with the application of computers allows one to work with models of very complex phenomena, undoubtedly beyond the limits of analytical or numerical methods. Simulation is the last resort technique in many technical sciences, but perhaps the most promising tool in the life and socio-economic sciences (the reader is referred to the Appendix for more detail on simulation).

The three main objectives of a numerical solution method for the modeling problem are:

1. Convergence to a solution and ability to terminate the computation,
2. Computational efficiency,
3. Accuracy, or error control.

However, other considerations may also be important, such as:

4. Ability to solve modified or extended formulations,
5. Existence of computer programs implementing a particular method,
6. Ease of use and conceptual clarity of the method.

Different measures of efficiency for numerical methods have been developed and the most common involve:

1. Counts of arithmetic operations needed to solve the problem,
2. Order of convergence for iterative processes.

The most practical measure of efficiency is the total computer resource needed for solution, including time and storage space. Unfortunately, it is frequently difficult to relate this measure to more theoretical properties of the model formulation and solution method. To reduce the total computer resource required for solution, we use a variety of mathematical techniques. Among them are:

1. Decomposition of large-scale problems into smaller components solved semi-independently
2. Linearization, whenever possible,

3. Data compression,

4. Reduction of problems with unknown degree of difficulty to well-known problems with which we have established experience.

Accuracy of the numerical solution depends mainly on quality of data, degree of model approximation, and numerical properties of the solution method. Each of these three factors can independently invalidate the solution results. It is fortunate that recent developments in numerical analysis have greatly contributed to our understanding of error analysis and conditioning of many numerical problems and algorithms. However, we remind the reader that in order for the model to be useful, modeling results must be interpretable.

To summarize, in this stage we obtain an analysis of problem solvability and a solution method. We should examine the solution method, its convergence properties and the error estimate. Such an examination may reveal that the modeling purpose is not achievable (e.g., because too much computer resource is required for modeling). In this case the modeler may wish to revise model definition and/or purpose. A pilot model may have to be developed to study these questions.

1.4.4 Computer Program Design, Development, and Checkout

This stage of the modeling process depends mainly on the size and complexity of the problem being implemented on the computer.

A mathematical modeling problem can seldom be completely solved by means of existing computer programs without any modification or additional programming effort. It is more often than not that we face the task of computer program design and development. This is accomplished in several stages, including:

1. Creation of complete and precise system of programs (software) corresponding to the mathematical model.

2. Adaptation of these programs to a selected computer or a set of computers (hardware) and operating systems.

3. Design and development of the pertinent data structures.

4. Program testing and debugging.

5. Documentation.

6. Application of a maintenance and change control procedure.

To design a complete and precise description of a large system of programs is a formidable task. Relatively small programs can be handled as one entity. A fundamental approach to large programs is their decomposition to smaller units to obtain manageable modules. The modules should have maximum independence; i.e., perform a well-defined function without complicated interactions with other modules in the program. They are essentially independent, functional, and logical components of a larger set. These properties of modules simplify their understanding, coding, and testing. In FORTRAN coded programs, for example, we have a main program and subprograms—subroutines and functions, which are the natural modules in the program. Other languages such as ALGOL permit nesting of independent procedures to an arbitrary depth.

Next we turn to a computer program library, if one is available, and attempt to find modules or "building block" subroutines. Most computer centers have libraries of subroutines for all the standard mathematical problems: solution of equations, optimization, quadrature, curve and surface fitting, and statistical computations. Model-building effectivity and efficiency can be considerably enhanced by use of such subroutines. Other computer program design considerations include: data organization, memory allocation and management and data input, input diagnostics, and solution output format.

Testing and debugging techniques are closely related to computer program design and documentation. If the system of programs consists of well-designed and documented modules the difficulty of testing and debugging can be substantially reduced. Programmers have also developed many practical approaches to debugging, such as snapshots of some intermediate and output values, post-mortem dumps, traces, and automatic debugging facilities provided by high language compilers. Neat documentation containing flow charts, indented code, and frequent comments greatly contribute to the minimum-error programming practices. Some efforts have been made to develop formal methods for proving correctness of computer programs. Limited results, applicable to analysis of computer program design, are discussed in Chapter 3 of this book. We remind the reader that "testing can show the presence of errors and not their absence" (Dijkstra). Moreover, computer programs usually change with time, making it imperative that their documentation be carefully maintained and updated. Programs should never contain outdated descriptions or data.

To summarize, in this stage the modeler has developed a tentative computer implementation of the model, ready for "fine tuning" and adjustment based on validation computations.

1.4.5 Model Validation, Adjustment, and Use

In many physical and engineering sciences, extensive past experience and a solid theoretical foundation help make validation a relatively simple matter. Ideally, checking out the model can be more or less reduced to verifying:

1. Accuracy of data and relationships, and comparison of modeling results with known results.

2. Numerical stability and accuracy of the computational process.

3. Computer program correctness.

4. Analysis of sensitivity of modeling results to changes in certain model parameters.

Usually a case or a set of cases can be tested and compared with known, theoretically or experimentally established results. Our confidence can be further strengthened by producing results comparable to those produced by other models and methods applied to the problem in question. If the results of all these tests are not satisfactory we have to either enrich the model or base it on a more accurate theory. Model enrichment is usually obtained by:

1. Including more details.

2. Adding variables and relationships between them.

3. Eliminating some restrictions and simplifying assumptions.

To switch to a more accurate theory, we may have to consider nonlinear relationships instead of linear, or stochastic elements instead of deterministic, etc.

Quite a different degree of difficulty may be involved in modeling socioeconomic, life, and management problems, where hypotheses are usually more speculative and data less accurate. Moreover, in many instances the model goals are not well defined and data for meaningful comparisons are not available. In these instances the model user should at least:

1. Assure himself that the model works the way he intended it to work.

2. Test to see if the model "predicts" known solutions using historical data.

3. Ask "experts in the field" to review the model structure, data, and results.

4. Check some extreme cases where solutions are easy to evaluate.

5. Keep records of the model usage and performance.

After the implemented model has successfully passed the validation stage, we can proceed to use it for modeling. Model adjustment and validation is, however, a continuing process: The modeling results will suggest further model adjustments and, once these are implemented, the model will have to be revalidated.

Different modeling experiments may have to be conducted to accomplish the modeling purpose. Each of these may represent one situation in prototype "space" (a design variation, an operating condition, etc.). Often several experiments are conducted in one computer "run," the only difference being in the setting of certain model parameters.

In using the model we should be aware of:

1. Limits of the model validity due to its structure, data, or both.

2. Range of parameters for which the model has been tested and results have been verified.

3. Possibility of model obsolescence due to new theories or data.

We should also bear in mind that our model was validated for a particular purpose and may, therefore, be invalid and useless for another purpose.

In a sense, the interpretation of modeling experiments is a reversal of the model-building process: To translate results from "model space" to "prototype space" the model user goes through all the modeling approximations in reverse. He takes into account computational errors, data inaccuracies, modeling approximations and discretizations, etc. Such proper interpretation of the modeling results is, of course, essential to the achievement of the modeling purpose.

REFERENCES

ALEXANDER, J. E., and J. M. BAILEY, *Systems Engineering Mathematics*. Englewood Cliffs, N.J.: Prentice-Hall, 1962.

ANDREWS, J. G. (ed.), *Mathematical Modelling*. London: Butterworths, 1976.

BAUER, F. L. (ed.), *Advanced Course on Software Engineering*. New York: Springer Verlag, 1973.

BENDER, E. A., *An Introduction to Mathematical Modeling*. New York: Wiley, 1978.

BONDER, S., "Operations Research Education: Some Requirements and Deficiencies," *Operations Research*, vol. 21, 797, 1973.

BREWER, G. D., *Politicians, Bureaucrats and the Consultant: A Critique of Urban Problem Solving*. New York: Basic Books, 1973.

BLACKWELL, V. A., *Mathematical Modeling of Physical Networks*. New York: Macmillan, 1968.

DAHL, O. J., E. W. DIJKSTRA, and C. A. R. HOARE, *Structured Programming*. New York: Academic Press, 1972.

DE SANTIS, R. M., and W. A. PORTER, "On Time-Related Properties of Nonlinear Systems," *SIAM Journal of Applied Math.*, vol. 24, 188, 1973.

EMSHOFF, J. R., and R. L. SISSON, *Design and Use of Computer Simulation Models*. London: Macmillan, 1970.

FURMAN, T. T. (ed.), *The Use of Computers in Engineering Design*. New York: Van Nostrand Reinhold, 1972.

GOLD H. J., *Mathematical Modeling of Biological Systems*. New York: Wiley, 1977.

HAWKES, N. (ed.), "International Seminar on Trends in Mathematical Modeling," Venice, 13–18 December 1971, *Lecture Notes in Economics and Mathematical Systems*, vol. 80. New York: Springer Verlag, 1973.

HYVÄRINEN, L., *Mathematical Modeling for Industrial Processes*. New York: Springer Verlag, 1970.

JACOBY, S. L. S., J. S. KOWALIK, and J. T. PIZZO, *Iterative Methods for Nonlinear Optimization Problems*. Englewood Cliffs, N.J.: Prentice Hall, 1972.

KWATNY, H. G., and V. E. MABLEKOS, *On the Modeling of Dynamic Processes*. Paper read at Engineering Foundations Conference: "Systems Engineering for Power," New England College, Henniken, N.H., 1975.

LEHMAN, R. S., *Computer Simulation and Modeling: An Introduction*. New York: Wiley, 1977.

MACHOL, R. E., *System Engineering Handbook*. New York: McGraw-Hill, 1965.

ROSENBLUETH, A., and N. WIENER, "The Role of Models in Science," *Philosophy of Science*, vol. 12, 316, 1945.

SHINNERS, S. M., *Control Systems Design*. New York: Wiley, 1964.

SMITH, C. L., R. W. PIKE, and P. W. MURRILL, *Formulation and Optimization of Mathematical Models*. Scranton, Pa.: International Textbook Co., 1970.

SMITH, J., *Computer Simulation Models*. New York: Hafner Publishing Co., 1968.

VEMURI, V., *Modeling of Complex Systems*. New York: Academic Press, 1978.

WILLIAMS, H. P., *Model Building in Mathematical Programming*. New York: Wiley, 1978.

WOLBERG, J. R., *Application of Computers to Engineering Analysis*. New York: McGraw-Hill, 1971.

Chapter 2

The Analytical Stage
of Model Building

This chapter deals with the analytical stage of model building, which includes the first three steps of the model-building process defined in Section 1.4:

1. Prototype identification and statement of the modeling problem, including definition of the modeling purpose and development of an overall modeling strategy.

2. Mathematical model definition and analysis.

3. Mathematical problem analysis, reformulation and solution development, including:

Definition and analysis of the mathematical modeling problem and its solvability, i.e., solution existence, uniqueness, stability, and conditioning.

Problem reformulation, if necessary and analysis of the error of approximation of the solution to the original problem by the solution to the reformulated problem.

Solution algorithm development and analysis of its properties of convergence, efficiency, and of stability.

In Section 2.1 we review the methodology involved in each of the above steps. This section is subdivided according to the above list (see Fig. 2.1). Subsection 2.1.5 deals with mathematical software that can be used as building blocks in computer implementations of solution algorithms for modeling problems. In Sections 2.2–2.4 the application of the methodology is illustrated in the development of three example models: a general, steady two-dimensional flow, a steady nonlinear pipe network flow, and a large-scale nonlinear optimization of a complex system operation. These examples have been selected because they illustrate various interesting aspects of mathematical modeling:

1. Both the two-dimensional flow and the nonlinear network flow models are useful for different modeling purposes and can be applied to a variety of prototypes. These examples are therefore of general interest as models.

2. Each of the three examples admits different modeling strategies; some of the alternatives are discussed.

3. One example is developed as a distributed and a lumped model, two are lumped models with particular structures: network and tree.

4. Each of the examples leads to a different mathematical modeling problem: linear and nonlinear algebraic equations and nonlinear optimization. In one of the examples, a differential equation is first derived and then reformulated as a difference equation problem. For another example we

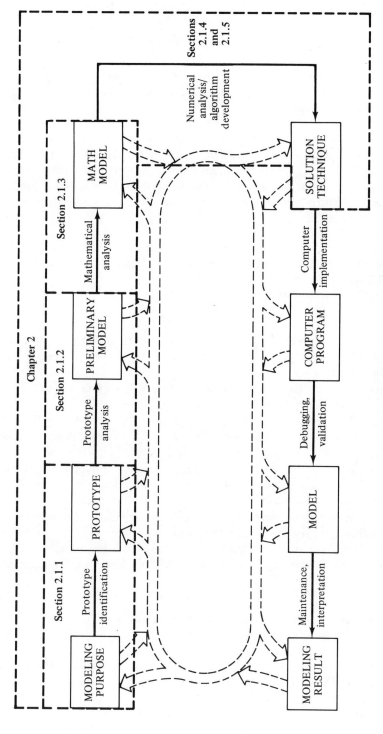

Figure 2.1 Guide to the discussion in Chapter 2.

37

present an algorithm for problem reformulation that can be implemented on a computer.

5. Each of the examples requires a different analysis and a different numerical solution algorithm and gives, therefore, rise to a variety of interesting mathematical analysis issues.

In Section 2.5 we provide a review in which the development of the example models is summarized. After the analytical stage of model development discussed in Chapter 2, the model is ready for computer implementation, the subject of Chapter 3. The steady nonlinear network flow model is used in Chapter 3 to illustrate the basic steps of the method proposed for the design of computer programs.

2.1 METHODOLOGY FOR THE ANALYTICAL STAGE OF MATHEMATICAL MODEL BUILDING

In this section we review the methodology for the analytical stage of the model-building process. The section is divided into subsections in which the topics have been grouped in sequential order for ease of presentation (see Fig. 2.1). This order is not meant to imply that models are built in this sequence of steps. The model builder in fact often cycles several times through different subsequences of these steps, each time refining his modeling approach. We present the motivation and outline the methodology for each of the analyses involved in model building. The objective of this presentation is to acquaint the reader with what has to be done and to develop the methodology to accomplish these tasks. A thorough theoretical treatise is not considered to be within the scope of this book. Several of the topics of the following subsections could be (some have been) dealt with in much more detail. For example, the development of efficient and stable numerical solution algorithms for the different mathematical modeling problems constitutes a subject for a book on numerical analysis. However, some of the considerations common to the construction of algorithms for most problems are discussed in this section.

2.1.1 Prototype Identification and Statement of the Modeling Problem

This subsection concerns prototype identification and statement of the modeling problem. Figure 2.1.1 shows the reader where we are in the model-building process.

Almost always, modeling is but one phase of a multiphased project. The

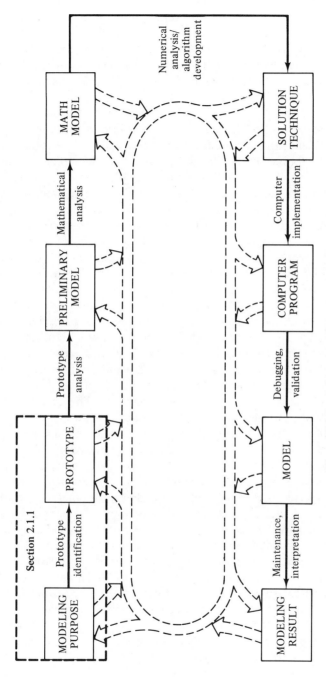

Figure 2.1.1 Guide to the discussion in Section 2.1.1.

modeler must not only understand the other project phases, but also support them; modeling for planning and design, for example, could be one phase of a major construction project. Modeling for observation, explanation or research could be a phase of a continuing effort to develop a scientific theory. Thus it is necessary, before the modeling project phase is launched, to define the modeling *purpose* in the general context of the overall project and to understand how to measure the result expected of the modeling phase in terms that are relevant to the overall project objective. In a sense then, the modeler must define the side conditions of the modeling project; for example, the "initial condition" is the starting point for model definition, and the "terminal condition" is the modeling purpose. Other side conditions include the information available for model derivation and use, as well as time and other resources available for model building and modeling.

Having delineated the modeling purpose in the above described sense, the modeler can proceed to *define the prototype* to be modeled and the questions, in "prototype space" he expects the modeling exercise to answer. A definition of the prototype includes a boundary, the definition of an environment that is relevant to the modeling questions, and the inputs and outputs that cross the boundary (if any do). This leads to a preliminary modeling approach: what kind of a model (or models) one might develop, and how it (they) could be used in the attempt to answer these questions. If possible, these candidate approaches should be ranked according to practicality, cost, and time required to achieve the modeling purpose. At this stage also a list should be developed of the information (type, detail, and format) necessary and available for model building and use. The definitions, statements, questions, and lists developed so far are essentially in the form of verbal statements, schematic drawings, etc. There is little if any resemblance to a mathematical problem statement. This will come later.

The information on this list and the preliminary modeling approach should be evaluated in the light of comments made in a paper by Koopman [1977]: The specific issue that should be addressed is whether the answers expected of the model are contained in the existing information. Koopman notes that the answer to a concrete and meaningful (modeling) question may be determinate, probabilistically determinate, or indeterminate. It is *determinate* if it is contained in the existing information. It is *probabilistically determinate* if probabilities of various answers are determinate, so that they may be calculated from the existing information. If the answer is *indeterminate*, i.e., neither determinate nor probabilistically determinate, then before proceeding with model definition the model builder should either revise the questions he expects the model to answer or modify his preliminary modeling approach and restore determinacy (see below).

Another issue to be considered now is whether the determinacy is a property of the model only. It is possible that answers provided by the modeling exercise are not interpretable, or are useless in "prototype space." In this case, the modeling exercise is useless because it still leaves us with the same doubt and indeterminancy that motivated the whole investigation in the first place. The model builder should proceed with model development only if the prototype is determinate with respect to the modeling approach. This means that there is sufficient evidence showing that modeling results can be interpreted in "prototype space." We remind the reader of the first necessary (and sufficient) condition for model usefulness developed in Section 1.2, the condition dealing with the replacement of a phenomenon in an unfamiliar field by one in a familiar field. Clearly this condition would be violated if the prototype is indeterminate with respect to the modeling approach.

Five methods for restoring determinancy are proposed by Koopman:

1. Increasing the information available for modeling.
2. Reducing the set of modeling questions or their extent.
3. Parametrizing the information.
4. Use of limiting arguments.
5. Sequential decision making.

The first two methods are obvious; the others require some elaboration. *Parametrization* is used if different possibilities with respect to missing information can be identified and precisely formulated. Then the model is exercised for each such possibility or parameter setting. This will usually lead to several answers to the modeling question, one corresponding to each parameter setting. Use of *limiting arguments* is an application of the method of parametrization in cases in which the maximum difference between the modeling answers corresponding to different parameter settings is not excessive. That implies that the modeling result is insensitive to the differences in parameter settings. This is also related to the notion of model behavior sensitivity to modeling data. It is helpful if such sensitivity analysis leads to the conclusion that model behavior, or the set of modeling answers, is insensitive to the maximum likely variation in parameter settings. *Sequential decision making* may be used when the body of information available for modeling varies (increases) during the modeling exercise itself. Then added information is available for use in subsequent stages of this exercise. A special case of this arises in multistage process modeling, when modeling results for one stage of a process lead to additional information which is used at the next stage, and so on.

We conclude this discussion with another point made by Koopman [1977] and related to the necessary and sufficient conditions for model usefulness: Model and prototype determinancy are just necessary and not sufficient for model usefulness. This is so because the fact that the information available together with the modeling approach contains the answers expected of the model does not imply that these answers are easy to find or even possible to find. We have *pseudoindeterminancy* in the former case and *operational indeterminancy* in the latter.

Now that we have dealt with the determinancy issue, we can formulate an *initial strategy* for the project. This is a statement of the approach that will lead to answer the questions in "prototype space," given the information, the resources and all other conditions that have been defined so far. Often at this early stage several candidate strategies can be formulated. Once this is accomplished, a preliminary plan should be developed for each such strategy. A plan consists of a list of project tasks, what each task would accomplish, its relation to other tasks and resources and time required to accomplish it. By project tasks we mean the different steps of model building, implementation and use.

The result of this step of model building is often in the form of a list of *tasks*, and times at which these tasks are to be accomplished (project *milestones*). Associated with this information is a resource matrix with one column for each task and one row for each type of resource expected to be used in the model building and modeling project and the units used for its measurement. The elements of the resource matrix are the amount of the particular resource (row) used for the task (column). It is helpful to consider how realistic this plan is, and whether the modeling objective still appears to be achievable within the limitations of, for example, time and budget. Another important part of the project plan is a list of the dependencies of each of the tasks either on results of previous tasks or on information from "outside" the project boundary. This task interdependence and task dependence on information is frequently cast in the form of a so-called PERT chart, implying the proper sequencing of project tasks. The information itself constitutes one of the sets of necessary conditions for project accomplishment.

To summarize, at this stage the model builder has defined:

1. A modeling purpose, questions the model will answer,
2. The prototype, its boundary and environment,
3. An initial modeling strategy,
4. A modeling project plan.

In Subsections 2.2.1, 2.3.1, and 2.4.1 we discuss the prototypes and modeling purposes for the examples of Sections 2.2, 2.3, and 2.4 respectively.

2.1.2 Mathematical Model Definition and Analysis

This subsection deals with mathematical model definition and analysis. Figure 2.1.2 shows the reader where we are in the model-building process.

The degree of *model detail* must be determined at least tentatively before the model elements are defined and the model is developed. The issue of

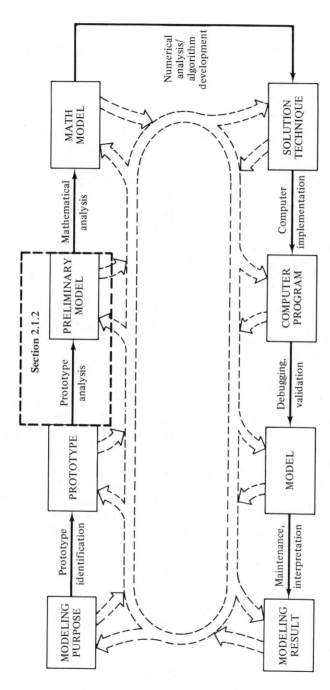

Figure 2.1.2 Guide to the discussion in Section 2.1.2.

43

model detail is one of trading off model realism versus modeling simplicity and practicality. Very detailed models require detailed prototype understanding and much data. They are expensive to build and to use. Aggregate, crude models, on the other hand, can be built on the basis of a general (or preliminary) understanding of the prototype. They may require less data for the modeling experiments and may be inexpensive to build and to use. Do detailed models always provide more detailed information about the prototype behavior? Sometimes they do, but in general this depends on several factors such as the amount and nature of computations involved in the modeling experiments. Probably more mathematical model building and modeling experiments fail to achieve their purpose because the degree of model refinement is poorly chosen than for any other reason. Factors involved in the determination of proper model detail include the prototype structure and behavior, information available for model development and use, and the modeling purpose. The question of model refinement should be considered in light of the necessary and sufficient conditions for model usefulness. If an improper level of detail (too little or too much) is selected, neither of these conditions may be satisfied; excessive computations in modeling experiments may be required, leading to uninterpretable, error-contaminated modeling results. Therefore, the first condition may be violated, the condition requiring that a phenomenon in an unfamiliar field (prototype "space") be replaced by one in a familiar field (model "space"). Too little model detail may not lead to meaningful modeling experiments at all. This would violate the condition that it should be possible to carry out modeling experiments under more favorable conditions than experiments with the prototype. Excessive model detail may also cause this condition to be violated, because it may require extensive and expensive model development and/or computations.

The most general case that needs to be considered is that of a complex prototype consisting of interacting components. Such a prototype is often represented by a *complex model* consisting of interconnected interacting submodels. One of the first steps in building such a model is to construct a schematic diagram, a simple iconic model. This simple descriptive model will suffice as a device for recording the next few model building steps. Although the overall model may be lumped (see Subsection 1.3.3 for definition), some of the component models may have to be distributed. In distributed models, an appropriate scale or resolution must be chosen for the spatial variables. In time dependent models, a time scale must be chosen. Modeling detail or resolution of the components must be balanced and matched with the modeling purpose. For example, if the modeling purpose is to develop an operational plan for a system, the time scale used in the model must be as fine as the desired plan (model output). If the desired plan consists of monthly average settings of controls, the model probably will not have to represent system behavior from one second to another, but rather weekly or monthly averages. Generally long-term planning models require less detail, i.e., a larger time scale and a coarser spatial resolution. On the other hand, detailed design models may require a fine spatial scale and a small time scale.

Modeling detail must be balanced with our degree of understanding of the prototype mechanism and the available modeling data. It may not make sense, for example, to represent the detailed spatial variation of a material property in the model if the only data available is an average value.

Another consideration in determining appropriate modeling detail is that of balancing model detail with the desired modeling accuracy and with data accuracy. If data and result accuracy is low, it may be sufficient to use a relatively crude representation in the model, say a lumped model approximating dependent variables on a rather coarse mesh.

Finally, there is the consideration of accuracy and amount of computation. Detailed models require more computations in the modeling exercise. Therefore, not only more computer resources would be used, but frequently it is also more difficult to control the computational accuracy and to interpret the modeling results.

One useful approach is to start with a relatively crude (lumped) model and to add detail later if needed. Detail can be added in the form of additional components. In distributed models, detail can be added simply by refining resolution of the space variable, i.e., the mesh over which differential equations are discretized. In complex models one may increase model detail by breaking one lumped-component model into more lumped models (modules) or by introducing a distributed model instead. Another method of adding model detail is to increase the number of variables, for example, replacing a fixed parameter by a variable. A large and complex modeling problem may also be carried out in a sequence of relatively simple steps in which the complex model is partitioned into a number of semi-independent smaller parts which can be analyzed, and exercised separately. The combined results then represent the behavior and properties of the prototype. Decomposition may be advantageous even if the modeling process requires many iterations involving submodel exercise, because the submodels are much smaller and simpler. For an illustration of decomposition applied to models of hierarchical (or multilevel) systems the reader is referred to Pearson [1966]. Decomposition methods for process models are discussed in Himmelblau [1973]. Applications of decomposition to static and dynamic optimization problems can be found in Lasdon and Schoeffler [1966], Schoeffler [1971], and Wismer [1971]. It is difficult to give a general methodology of decomposition because the specific problem properties, such as structure and sparsity of matrices, usually are utilized for model formulation and reformulation.

To define the *model elements*, use is made of laws of physics such as balances of quantities in "control volumes," of approximation, rules of behavior, statistical information and of systems analysis. Systems analysis is useful in determining the proper model structure. In modeling of complex prototypes it is useful to start by defining the model boundary, its component boundaries and constructing an *interaction matrix*. This matrix has a row and a column for each prototype component, and its elements indicate the interactions or interfaces between the components. Model components can be functionally, mechanically, structurally, logically, or otherwise linked. An example of logical linkage is a set of events that can occur only in sequence. Functional linkage simply means that the output of one component is the input to another component. Interaction of prototype elements with the environment can also be recorded on the interaction matrix. In a sense then, the information in this matrix implies sets of inputs and outputs of the components.

Next, the *variables* and *parameters* in each of the component models, and the mathematical and logical *expressions* relating them must be defined. Two types of independent variables appear in distributed models: spatial coordinates and time. Accordingly these models can have one, two, or three spatial dimensions and a fourth, or time "dimension." Lumped models do not have spatial variables, but time may appear as an independent variable in these models. We also distinguish two types of dependent variables, across and through variables:

1. *Across variables* (denoted below by p, for potential) are scalars relating the condition at one point to that at some other point. The across variables are usually measured as a (potential) difference between values at two points in the model or between the value at one point in the model and at one reference point such as ground. Examples of across variables include velocity, voltage, pressure, and temperature in engineering models and price differentials in economic models.

2. *Through variables* (denoted below by f for flux) are vectors representing the flux of some quantity traversing an elemental cross section of the model. Examples of through variables include force, torque, current, fluid and heat flow in engineering models and commodity flow in economic models.

Two quantities related to across and through variables are power and energy, defined as follows:

1. *Power P(t)* is the product of the across and through variables:

$$P(t) = p(t) \cdot f(t)$$

2. *Energy $E(t)$* is the integral with time of power from $-\infty$ to t:

$$E(t) = \int_{-\infty}^{t} P(\tau)\, d\tau$$

$$= \int_{t_0}^{t} P(\tau)\, d\tau + E(t_0)$$

where $E(-\infty) = 0$ and $E(t_0)$ is the energy at time t_0.

A model is said to be an energy *sink* or a *load* if its energy for all t is non-negative. Otherwise it may be a *source* or a *driver*.

Mathematical and logical expressions relating across and through variables, other variables, and parameters must now be assembled or developed. These expressions represent the laws known to apply to the prototype. They could be physical laws; relations representing the mechanism by which component output is derived from input; or any other symbolic representation of our knowledge (or hypotheses) about the prototype. Often component models, or specifications from which such models can be derived are available. Component models for many engineering systems, for example, are supplied by their prototype manufacturers. If they are not available, component models have to be developed. This can be accomplished with analytical methods or with empirical methods. Examples of the latter would be laboratory measurements to determine component output as a function of component input, or statistically designed computer experiments to determine component responses to inputs.

Mathematical and logical expressions in the component models must be valid over the whole range of variation of the independent parameters in the modeling experiments. If this is impossible, modeling experiments must be restricted to areas where these expressions are valid. We should note at this point that iterative procedures are often used to solve the mathematical modeling problem. In such procedures, sequences of trial solutions are evaluated, leading to simultaneous satisfaction of all modeling relationships. Some of these mathematical trial solutions may lie outside of the normal (i.e., physical or real-life) ranges of variation of the independent variables. It is important that the model relationships and all parameters entering these relationships be valid for any values of the independent variables likely to occur during modeling solutions. This may be broader than the range of physically meaningful or realizable values.

One analytical approach to component model definition is further described in the Appendix on *simulation*. Some of these models essentially keep track of the state of the components as represented by the independent variables, and of state changes. Component inputs, outputs, and interaction are as implied by the component interaction matrix. The mechanism which causes these changes in the prototype component is normally not represented in the logical simulation model. During the modeling experiments, these state changes are externally triggered by the model user by means of inputs. Frequently these triggering inputs are nondeterministic, that is, statistically derived.

Another approach to analytical definition of component models uses "*control*" systems models. We use quotation marks because this approach is not limited to control systems as such. It is frequently applied to a variety of prototype systems and processes. The approach originates with control systems design and studies of process stability but has broader application. In these prototypes (systems or processes) one is interested in keeping track of the state of certain variables and in using this information to achieve and maintain a (given) desired state. The method used to accomplish this in the prototype is to compare the actual state to the desired state and to use the difference (if any exists) to trigger operational changes which will eliminate this difference. Such systems or processes and their models usually consist of several components which operate on certain inputs and generate outputs. As with any other complex prototype and model these components are functionally interconnected such that outputs of certain components are inputs of certain other components. These interconnections are again determined and recorded in the previously described interaction matrix. Component model definition, then, consists of a mathematical statement of the law (e.g., physical) applicable to generation of the component output(s) as a function of the input(s). These could be simple statements, such as a sum, or complicated functional relationships. Smith [1977] gives a comprehensive review of mathematical modeling techniques applicable to continuous and discrete processes.

One of the more commonly used methods for analytical derivation of component models is based on applications of the principle of *conservation* of certain quantities in the model. The conservation principle considers the boundary and the interior of the component model. The resulting mathematical model is simply a *balance equation* for the conserved quantity. It states that the net input which crosses the boundary, plus generation within the boundary, equals accumulation (or depletion) within the boundary. Examples of conserved quantities that can be used in this modeling approach include: mass moment of inertia, heat, force, torque, products of chemical reactions, dollars, and many others. In lumped (zero-dimensional) models the boundary is the component boundary and the interior is the component interior. In distributed models the conservation principle is applied to a "control space" of infinitesimal size. This space is a volume if there are three independent spatial variables, or an area if there are two.

Smith et al. [1970] apply the conservation principle in the derivation of a variety of models. The basic equation is of the form:

Input − output + generation = Accumulation − depletion,

where all terms refer to the conserved quantity. Often it is more convenient to deal with rates, and the equation becomes:

Rate of (input − output + generation) = Rate of
(accumulation − depletion),

where again all terms refer to the conserved quantity. The terms input, output, etc., entering these equations are to be understood in a general sense. For example, the conserved quantity could be momentum. The component model would take the net rate of momentum, plus the sum of the forces acting on the component and equate this to the product of acceleration and mass within the component. In the (special) case of a steady-state model, acceleration is zero and, therefore, the net rate of momentum plus the sum of forces equals zero.

In models of chemical reactions, products of the reaction may appear in the control space (usually control volume in this case). Application of the conservation principle will yield the balance equation:

Rate of (input + appearance) = Rate of
(output + accumulation),

where all terms refer to a particular component of the chemical reaction (the conserved quantity).

One balance equation must be formulated for each conserved quantity (or, e.g., each chemical component) in each component model. There may be more than one conserved quantity in each component.

Other information may be available about the prototype operation, in addition to the fact that certain quantities are conserved in it. This information may be in the form of laws defining physical phenomena. Mathematical expressions of these laws in terms of the model variables should be derived and combined with the balance equations. Frequently these "laws" can be obtained by application of a variational principle (see Subsection 2.2.5 for an illustration). It consists of stating the "law" in terms of the minimization of an integral. This approach was first developed for applied mechanics problems, and then carried over to electromagnetics, fluid mechanics, and other areas. This approach can lead to useful analytical results and also to convenient computational schemes (see Subsection 2.2.5).

Three basic types of physical phenomena or laws in prototype space may be represented in mathematical models. These representations are basic elements of distributed or lumped models, also referred to as element characteristic functions, relating the model variables and parameters. These model elements also have an orientation. In distributed models, the orientation is that of the (set of) spatial coordinate(s). Lumped-model elements have an initial (input) node and a terminal (output) node. The three types of model elements are (Smith et al. [1970]):

1. Dissipators,
2. Reservoirs of flux,
3. Reservoirs of potential.

Dissipators conduct a flux of a magnitude that is proportional to a potential gradient, or difference across the element. The factor in the proportion is referred to as *dissipativity* (or its inverse—*conductivity*) per unit of length. This factor can be a fixed parameter or a variable that depends on position. In dissipators energy is dissipated or, equivalently, the entropy of the element and its environment is increasing. For this reason lumped dissipator elements are also referred to as *lossy* elements. The dissipation of energy is related to the flux through the element. Denoting the independent variable (spatial coordinate) by x, the through variable (flux) by f, the across variable (potential) by p, and the dissipativity per unit length in the x-direction by D_x, the distributed model of a dissipator is

$$D_x f = -\frac{\partial p}{\partial x}$$

and the lumped model of a dissipator is

$$Df = -\Delta p_{12}$$

In the latter model Δ denotes a difference and the subscripts 1 and 2 denote the initial and terminal nodes at which the across variable is measured.

The power of a lumped dissipator is given by

$$P(t) = \Delta p(t) \cdot f(t) = \frac{1}{D_x}[\Delta p(t)]^2$$

which is obviously positive, and the energy is given by

$$E(t) = \int_0^t P(\tau)\, d\tau + E(0) = \int_0^t \frac{1}{D_x}[\Delta p(\tau)]^2\, d\tau + E(0)$$

which is nonnegative over any time period. As stated, systems with this property are referred to as loads or sinks into which energy flows.

An example of a distributed model of a dissipator is the electric conductor:

$$RI = -\frac{\partial v}{\partial x}$$

in which v denotes voltage (across variable), I denotes current (through variable) and R the resistance (dissipativity). An example of a lumped model of a dissipator is the flow of liquid through a pipe section

$$cq^\alpha = -\Delta p_{12}$$

where Δp_{12} denotes the difference between the pressures at the two pipe section ends (across variable), q denotes the liquid flow (through variable) and c and α are constants which depend on pipe geometry and fluid viscosity (dissipativity).

In *reservoirs of flux* (or of energy) the rate of change of flux with respect to time is proportional to the potential gradient or difference across the element. The factor in the proportion measures the capacity of the flux reservoir per unit length in the x-direction. It may be a fixed parameter or a variable that depends on position. Denoting this parameter by C_{f_x} and time by t the distributed model of a reservoir of flux is:

$$C_{f_x}\frac{\partial f}{\partial t} = -\frac{\partial p}{\partial x}$$

and the lumped model is

$$C_f \frac{\partial f}{\partial t} = -\Delta p_{12}$$

The power of a lumped reservoir of energy is:

$$P(t) = \Delta p(t)f(t) = -C_{f_x}f(t)\frac{\partial f(t)}{\partial t}$$

which in general can be either positive or negative, as can the energy. It can, however, be shown that for any period of time for which the initial energy is zero, the energy of an energy reservoir cannot be negative. Under these circumstances the energy reservoir cannot act as an energy source.

An example of a distributed model of a reservoir of flux is the fluid flow:

$$\frac{p}{g}\frac{\partial v}{\partial t} = -\frac{\partial p}{\partial x}$$

where v is the velocity of flow in the x-direction, ρ the fluid density and g the gravitational constant.

In *reservoirs of potential* the rate of change of flux with respect to distance is proportional to rate of change of potential with respect to time. The factor in the proportion measures the capacity of the potential reservoir per unit length in the x-direction. It may be a fixed parameter or a variable that depends on position. Denoting this parameter by C_{p_x} the distributed model of a reservoir of potential is:

$$C_{p_x} \frac{\partial p}{\partial t} = -\frac{\partial f}{\partial x}$$

and the lumped model of a reservoir of potential is

$$Cp \frac{\partial \Delta p_{12}}{\partial t} = -\Delta f_{12}$$

Examples of reservoirs of potential include capacitance elements of electrical systems, and heat conducting elements with heat storage capacity. The power of a lumped reservoir of potential is:

$$P(t) = \Delta p(t) \, \Delta f(t) = -C_{p_x} \, \Delta p(t) \frac{\partial \Delta p(t)}{\partial t}$$

which can be either positive or negative, as can the energy. It can be shown that for any period of time for which the initial energy is zero, the energy of a reservoir of potential cannot be negative. Under these circumstances the potential reservoir cannot act as an energy source.

Further examples of prototype system elements and corresponding model elements are listed in Table 2.1.1 (compare Blackwell [1968] and Smith et al. [1970]).

The conservation principle method to component model definition can be summarized as follows:

1. Define: control space or system and boundary, quantities conserved, independent and dependent variables and parameters.

2. Define: input, output, accumulation, depletion and generation, and balance equation for each conserved quantity.

3. Develop mathematical statement of laws applicable to input, output, etc. defined in item (2) in terms of variables, etc. defined in (1). It should be noted that these "laws" are often established empirically.

4. Combine laws with a balance equation for each conserved quantity.

5. Record all simplifying assumptions, limitations on validity of data,

Table 2.1.1 Examples of Prototype Systems and Corresponding Model Elements

PROTOTYPE SYSTEM	ACROSS VARIABLE	THROUGH VARIABLE	DISSIPATORS		RESERVOIRS OF FLUX		RESERVOIRS OF POTENTIAL	
			RESISTANCE PROTOTYPE	DISSIPATIVITY	INDUCTANCE PROTOTYPE	FLUX RESERVOIR CAPACITY	CAPACITANCE PROTOTYPE	POTENTIAL RESERVOIR CAPACITY
Translational mechanical	Translational displacement/velocity	Force	Translational damper	Viscous damping	Translational spring	$(\text{Spring constant})^{-1}$	Translational mass	Mass (inertia)
Rotational mechanical	Angular displacement/velocity	Torque	Rotational damper	Viscous damping	Rotational spring	$(\text{Spring constant})^{-1}$	Rotational mass	Inertia
Electrodynamic	Voltage	Current	Electrical resistance	$(\text{Resistivity})^{-1}$	Inductance	Inductivity	Electrical capacitance	Capacity
Fluid dynamic	Pressure (velocity potential)	Fluid flux rate	Fluid resistance	$(\text{Viscosity})^{-1}$	Fluid inertance	Inertia (density)	Fluid capacitance	Compressibility
Heat transfer	Temperature	Heat flux rate	Thermal resistance	$(\text{Thermal resistivity})^{-1}$	–	–	Thermal capacitance	Thermal capacivity

laws, etc. This information is an essential part of the model as it may limit its applicability or validity.

6. Repeat (1)–(5) for all component models.

The next step is that of component interaction model definition. This is essentially a process of developing balance equations for all component inputs and outputs, and the overall model boundary conditions. For this purpose, we use the inputs and outputs of the model components and the previously defined interaction matrix. The conservation principle is applied to each conserved quantity throughout the model and across its boundaries. Component models which may be distributed or lumped, are interconnected at common nodes implied by the interaction matrix. Each component model has input (initial) and output (terminal) nodes for each conserved quantity. A *path* is said to exist between two nodes if it is possible to trace a line from one to the other without losing contact with interconnected components. A *circuit* (or a *loop* or *mesh*) is a set of interconnected components which form one and only one closed path (i.e., a path whose initial and terminal nodes are one and the same). The relationships between through and across variables in complex models are referred to as the two *Kirchhoff laws* of network flow:

1. The *first Kirchhoff law* is the *node law*; it states that the oriented sum of the through variables at every node is zero. The *oriented sum* of the through variables at a node assigns one sign (say $+$) when the variable is oriented away from the node and the opposite sign (say $-$) when the variable is oriented toward the node.

2. The *second Kirchhoff law* is the *circuit law* or *loop* law; it states that the oriented sum of the across variables around every circuit of interconnected components is zero. An *oriented sum* of across variables around a circuit takes into account a given circuit orientation. It adds across variables whose orientation is the same as that of the circuit and subtracts across variables whose orientation is opposite to that of the circuit. This implies that if two components are interconnected, the across variable can only assume one value at the interconnection.

The two Kirchhoff laws at the nodes and on the circuits of the complex model are represented in the model in terms of algebraic equations. For a model with m nodes and ℓ components, the first Kirchhoff law yields $m - \ell$ independent equations and the second Kirchhoff law yields $\ell - m + 1$ additional independent equations. This subject is further explored in the discussion of a network flow example in Section 2.3. Examples of Kirchhoff laws in different models include (compare Blackwell [1968] and Smith et al. [1970]):

PROTOTYPE·	MODEL NODE LAW	MODEL LOOP LAW
General complex system	Vector sum of components' inputs/outputs of each conserved quantity	Algebraic sum of components' across variables for each conserved quantity
Electric network	Algebraic sum of currents	Algebraic sum of voltage differences
Hydraulic network	Algebraic sum of flows	Algebraic sum of pressure differences
Translational mechanical system	Vector sum of forces	Algebraic sum of displacements/ velocities
Rotational mechanical system	Algebraic sum of torques	Algebraic sum of angular displacements/velocities
Transportation system	Algebraic sum of transported commodities	Algebraic sum of price differences

The final step in model definition is that of formulation of *side conditions*: boundary and initial conditions. Boundary conditions are model elements representing inputs or outputs of model components from or into the environment respectively. These are defined in a fashion analogous to the definition of inputs or outputs from one model component to the other. If the model is not steady-state, initial and/or terminal conditions must be defined. These will define the state of the model at the beginning or the end of the modeling exercise respectively. It should be noted that the conservation principle is applicable to the model as a whole, just as it is to each of its elements. Therefore, the sum of all boundary conditions should represent a balance equation for all quantities that are conserved in the model. If the model is not steady-state, this balance is applicable to the total period of model exercise.

The steps of model development discussed in this subsection can be summarized as follows:

1. Determine model detail, component breakdown, and structure.

2. Define component model interaction matrix and record in it all interactions, component inputs and outputs including interactions of components with the environment and boundary conditions. If the model is not steady-state, define initial and/or terminal conditions.

3. Develop component models and, using the interaction matrix developed in step (2), balance all component inputs and outputs and the model boundary conditions.

4. Summarize and record all assumptions and restrictions that were made in steps (1)–(3) and may limit model validity.

The steps described in the foregoing discussion are illustrated in Subsections 2.2.2, 2.3.2, and 2.4.2, in which we develop mathematical models for the examples of Sections 2.2, 2.3, and 2.4 respectively. These steps result in a preliminary model definition, which in turn constitutes a record of the model builder's knowledge of the prototype, its structure, elements, properties, attributes and its operating mechanism. Only those aspects are represented which the model builder considers relevant to the modeling objective. Also, degree of approximation, detail, and resolution are as deemed appropriate for the modeling purpose. This record is a preliminary model because it may not be a mathematically or logically complete and consistent set of relationships that leads to a mathematically solvable modeling problem.

The preliminary model definition consists of a set of mathematical relationships. Distributed-model elements involve infinitesimal variations in a continuum and lead to differential equation models. Depending on the number of independent variables (spatial coordinates or degrees of freedom) these could be ordinary or partial differential equations. Unsteady models, distributed or lumped, involve derivatives of dependent variables with respect to time, and therefore also lead to differential equations. Ordinary differential equations (or equation systems) occur in:

1. Distributed steady-state models with one independent variable (spatial coordinate or degree of freedom) in each model element.

2. Unsteady-state models with time being the only independent variable.

Partial differential equations occur in models with more than one independent variable (spatial coordinate or degree of freedom, or time). Normally only a small number of different differential equations appear. Models with one type of element only, i.e., dissipator or reservoir of flux or reservoir of potential, will lead to elliptic partial differential equations (Laplace's or Poisson's equation). Models with dissipator and reservoir of potential or reservoir of flux (but not both) elements lead to parabolic partial differential equations (diffusion equation). Models with a reservoir of flux and of potential but no dissipator lead to hyperbolic partial differential equations (wave equation).

Discretized reformulations of ordinary and partial differential equation models lead to algebraic equations, as do many lumped models. Also, all models of the Kirchhoff laws are algebraic equations. The model definition may have to be modified to obtain a conveniently solvable modeling problem.

We remind the reader that in order to be useful, models should satisfy two necessary and sufficient conditions (Rosenblueth and Wiener [1945]):

1. The model builder-user must be more familiar with the mathematical problem involved in studying model behavior than with the problem of

studying prototype behavior directly; a phenomenon in an unfamiliar field is thereby replaced by a phenomenon in a familiar field.

2. It must be possible to carry out modeling experiments under more favorable conditions than would be possible when experimenting with the prototype itself. This necessary condition implies that the mathematical model must be based on an adequate theoretical model which leads to nontrivial modeling experiments. Modeling experiments with a mathematical model can be conducted only if the mathematical problem involved in the modeling experiments is mathematically solvable and if a solution to it can be computed if necessary. (To satisfy the latter condition, we may need a convergent and usable algorithm.)

The analysis of problem solvability is the subject of Subsection 2.1.3, the development of solution techniques is discussed in Subsection 2.1.4.

2.1.3 Mathematical Modeling Problem Analysis and Reformulation

The activities described in the previous subsections produce a tentative definition of a mathematical model. This model consists of mathematical relations involving variables, parameters, data, etc., and constitutes a mathematical problem statement. The mathematical problem could, for example, be a set of equations or a set of inequalities and an associated optimization objective. To achieve the modeling purpose, the mathematical modeling problem must be solved. Frequently repeated solutions of the same basic problem are required subject to different data, parameters, or side conditions. Each of these problem variants corresponds to a different case study, or to a different set of circumstances determining prototype behavior. At this stage then, the modeler should be concerned with an overall solution strategy for the mathematical modeling problem: given the available data and model definition, obtain a solvable problem, develop a solution method, implement the method on a computer (if necessary), and solve the problem or set of problems to achieve the modeling purpose. This strategy must be valid for all variations of the modeling problem that may arise, for all data, parameter values, etc. Therefore, the model definition and the mathematical problem implied by it must remain the same for all modeling studies to be performed. This is a necessary condition for the solution method to be applicable to all modeling studies which will have to be performed. Also, in developing the solution method, the modeler should insure that all elements of the method remain valid and sufficiently accurate for all data and parameter values that may arise in the modeling studies.

Once a tentative or preliminary model definition is developed, its properties must be analyzed and candidate solution techniques must be selected.

The processes of problem formulation and solution method development are interrelated and frequently accomplished in parallel. Availability of a good solution technique may dictate the modeling approach to a given problem. Sometimes a new solution technique has to be designed and developed to deal with an unusual modeling problem that cannot be reduced to a standard mathematical problem and be handled by an available method. In our discussion we will separate issues related to problem formulation from those related to solution methodology. The main questions in the first category which will be dealt with in this subsection are those of solution existence, uniqueness, and stability, and problem reformulations. In the following subsection we have grouped the issues of algorithm development, conditioning, numerical stability, computational efficiency and error analysis. The discussion of computer implementation is presented in Chapter 3. Figure 2.1.3 shows where we are in terms of the model-building process.

The first question is that of *solution existence*: Does the mathematical modeling problem have a finite and mathematically determined solution? It is appropriate to resolve this question at this stage of the development, for if the problem is not solvable, the modeler should obviously not proceed to attempt to solve it. Instead, he should adjust the model definition so that a solvable problem is obtained and the modeling purpose can be achieved. A mathematical problem may not be solvable because it is overdetermined, i.e., different parts of the problem statement, such as equations or side conditions, are redundant or contradictory. The question of solution existence then is: Does the mathematical problem statement properly determine a solution? This question is often resolved by construction of a solution; if a solution to the problem can be constructed and shown to exist in general, then solution existence is established. This approach is frequently used to establish existence of solutions to initial or boundary value problems involving differential equations (ordinary or partial). Another approach is to formulate an equivalent problem which is known to have a solution, thereby establishing existence of a solution to the original problem. We illustrate this approach in Section 2.3.

Many analyses have been published proving either existence or nonexistence of solutions to certain problems. In some cases, nonexistence has been shown to result from introduction of "simplifying" assumptions into the problem statement. Thus, it is possible to state a mathematical modeling problem that has no solution in general. Often a solution existence investigation cannot be pursued for practical reasons such as the problem's being too complex, as in the case study described in Section 2.4; in this case, a mathematical programming problem is defined, that is: find a set of vectors $\mathbf{x}, \mathbf{y}$ minimizing an objective function $F(\mathbf{x}, \mathbf{y})$ and satisfying a set of constraints. Some of these constraints are constants and others are functions of the variables $\mathbf{x}, \mathbf{y}$. A pair of $\mathbf{x}, \mathbf{y}$ satisfying these constraints is called feasible.

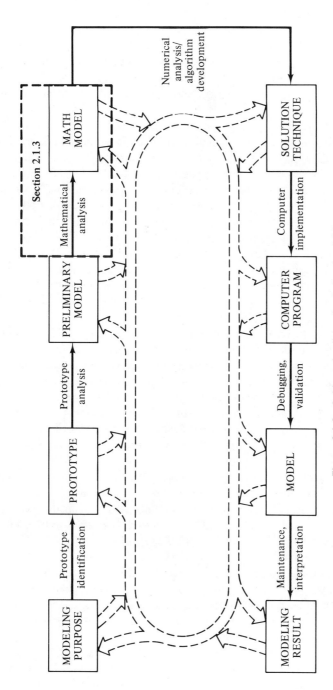

Figure 2.1.3 Guide to the discussion in Section 2.1.3.

Hence, we are looking for a feasible pair of **x**, **y** that minimizes $F(\mathbf{x}, \mathbf{y})$. This set of variables and constraints represents an existing physical system. It has been designed by engineers who selected a set of design conditions that should be satisfied simultaneously. In the mathematical model, however, the number of variables and constraints is large and it is impossible to determine in general that corresponding to any set of data there is a feasible pair of **x**, **y**. Moreover, experience with the prototype described in Section 2.4 indicates that some design conditions (constraints) have always been violated to some extent for all **x**, **y**. Clearly, we have to reassess our requirement of strict feasibility and settle for a more practical goal of finding a minimum of $F(\mathbf{x}, \mathbf{y})$ subject to "tolerable violation" of certain constraints. This compromise was the only way to remove the problem of solution nonexistence inherent in the case. Because of this compromise, an algorithm had to be found that would generate a sequence of approximations to a minimum $(\mathbf{x}^*, \mathbf{y}^*)$, without requiring a strictly feasible initial approximation $(\mathbf{x}^0, \mathbf{y}^0)$. In other mathematical modeling problems such a compromise may not be possible. It would then be necessary to modify the model or the data and obtain an acceptable problem formulation for which a solution can be shown to exist, at least in theory.

It is useful to recall that boundary and initial value problem statements must include boundary and initial conditions. These conditions specify either the dependent variable or a derivative of it for some given value of one independent variable and every value of the other independent variables. Depending on whether or not derivatives are specified, these conditions are either differential or algebraic equations. The number of side conditions needed for a problem to be solvable, depends on the order of the derivatives and on the number of independent variables. The required number of boundary conditions, for each dependent variable in terms of one of the independent variables, equals the order of the highest derivative with respect to that independent variable. Thus, elliptic partial differential equations such as Laplace's equation for example have second-order partial derivatives with respect to each independent variable. Therefore, they require two boundary conditions for each independent variable, specifying either the dependent variable or its derivative at the boundary.

The next question to be asked concerns *solution uniqueness*. If the problem has many, possibly infinitely many solutions, the modeling purpose may not be achievable. Say, for example, that the modeling purpose is to study the performance of a proposed engineering design. Several (possibly infinitely many) different prototype performances may be implied by the model, for each set of conditions if the modeling problem does not have a unique solution. If this is not acceptable, the modeling problem must be reformulated such that it can be shown to possess a unique solution. The examples described in Sections 2.2 and 2.3 illustrate cases in which solution uniqueness is essential. To study and establish uniqueness, one usually assumes the existence of two distinct solutions to the stated problem. These distinct solutions must satisfy all conditions of the problem statement including initial and boundary data. Solution uniqueness is established if it can be shown that under the given conditions the two (and hence any two) solutions which were assumed to be distinct are indeed identical. This analysis is illustrated for the boundary value problems defined in Sections 2.2 (partial differential equation) and 2.3 (nonlinear algebraic and transcendental equation set). In many practical problems, it is impossible to determine whether or not the mathematical modeling problem has a unique solution. The example described in Section 2.4 is again a case in point; in mathematical optimization, there is a theorem stating roughly that a strictly convex objective function has a unique minimum over a convex feasible region (subject to some additional qualifications). In many cases, such as the one under consideration in Section 2.4, it is virtually impossible to establish whether the problem satisfies these requirements. We usually settle for less, which is a local minimum (one of many possible). This is an acceptable approach to many optimization problems; if our iterative optimization algorithm generates a better solution than a given starting point (if not a local minimum) than we can claim that the model was useful and the modeling purpose has been achieved.

The question of *solution stability* is whether the solution depends continuously on data such as boundary conditions for example. This is an important factor, especially where modeling data may be inaccurate. The inaccuracy could be the result of measurement errors, if the data is obtained by measurement of experimental results, for example. Another source of data inaccuracies is computational error; data used in a model could be the result of inaccurate computations involved in another model. Before proceeding to use possibly erroneous data to solve a modeling problem, we should make sure that the problem is stable. The property of solution stability, i.e., of continuous dependence of the solution on the side conditions is necessary so that small data errors may not result in large solution errors. The method of analysis used to determine whether a problem possesses this property is similar to solution uniqueness analysis: we assume the existence of two distinct solutions to the stated problem, corresponding to and satisfying two sets of (e.g. boundary) data which differ by a small amount. In general, the solution is stable if it can be shown that these two (and hence any two) solutions which were assumed to be distinct differ by a small amount anywhere in the region of interest.

The question of *problem conditioning* has to do with sensitivity of the problem solution to perturbations in problem data. This issue is best addressed during the development of a solution technique, and should be examined again in the model validation stage. Accordingly, we discuss problem conditioning in Subsections 2.1.4 and 4.3.1 which deal with these topics.

Often *problem reformulation* is a required part of the overall solution strategy. If for example the mathematical model is a set of differential equations, the modeling problem is a boundary or initial value problem involving these equations. In most cases, this problem would have to be solved on a digital computer, often an impossible task without reformulation. To achieve the modeling purpose, the original problem will have to be reformulated and cast in terms of a set of algebraic equations that can be solved on a computer. Such a reformulation is described in Section 2.2. Often problem reformulation is motivated by the desire to increase efficiency and/or accuracy of the computation of problem solutions. An example of a reformulation is described in Section 2.3, which results in a problem of smaller dimensions (fewer independent variables and equations). This normally improves computational efficiency for nonlinear equation solution, the problem under consideration in this case. (Recent developments in sparse matrix solution technology show that this may no longer be a valid claim.) In other cases problem reformulation leads to a simpler problem, or to one for which standard solution techniques are available.

We should be cautious with problem reformulations because they may destroy problem structure or other properties which are essential for successful modeling. The importance of the original problem structure is that it may resemble the prototype structure whereas the reformulated problem structure may not. If the original problem is solved, model behavior may therefore yield more information about prototype behavior than would the solution of a reformulated problem. In the example of Section 2.4 we use a penalty function reformulation which in effect changes the problem structure completely. This is based on the premise that in the limit the solution of the reformulated problem tends to that of the original problem. Other problem properties which may be destroyed by reformulation are solution existence, uniqueness, stability, and conditioning. In fact, if problem reformulation constitutes an approximation, the modeler should reanalyze the reformulated problem to ascertain that it has these properties. In general, the existence of a unique solution to an approximation of a mathematical model is not guaranteed by the existence of a unique solution to the original, approximated problem. Furthermore, it needs to be shown that the solution of such a problem reformulation can be made to approximate the original problem solution as closely as desired. This means that the approximating problem solution must tend to the solution of the approximated problem as the approximation itself is made more accurate in some sense (e.g., by reducing mesh size). Such an approach is described in the case study illustrated in Section 2.2.

Another type of model reformulation is a change of some or all of the problem variables. This is done mainly for computational convenience. In many engineering models dimensionless normalized variables are obtained by subtracting from the original problem variable a reference (or characteristic) value of that variable and dividing the difference by a reference (or characteristic) difference. The benefit of such a change is that it makes the problem solution more general and applicable to the complete range of variation of this variable. Often changes of variable reduce the number of independent variables required for the mathematical model. The use of dimensionless parameters grouping several independent variables in fluid flow or heat transfer models are examples of this. To obtain these reformulations, we first change the original variables into dimensionless normalized variables as described above. These new variables are then introduced into all the model relationships. The relationships are then rearranged so as to obtain dimensionless groups or parameters. Examples can be found in fluid mechanics or heat transfer books or in Chapter 9 of Smith et al. [1970]. Often changes of problem variables are required to properly scale the numbers involved in computing problem solutions. This numerical analysis issue is important because of the limited computer accuracy and the vast number of arithmetic operations required to solve many problems. More about this subject can be found in the next subsection dealing with solution methods.

To summarize, at this stage the model builder has analyzed the mathematical modeling problem with respect to solvability and may have reformulated it to obtain a more convenient solution. In Subsections 2.2.3, 2.3.3 and 2.4.3–5 we discuss the mathematical modeling problem analysis for the examples of Sections 2.2, 2.3, and 2.4 respectively. Our mathematical modeling problem is now ready for a solution algorithm, the subject of the next subsection.

2.1.4 Solution Algorithm Development

At this stage of the model-building process (see Fig. 2.1.4), we have defined a model, a modeling problem and an overall solution strategy. We may have also determined that the modeling problem is theoretically solvable and has a unique and stable solution. We have, if necessary, reformulated the problem into a form that is more convenient to handle and is computationally feasible. This subsection deals with the development of solution methods or algorithms for numerical problems. (Logical simulation models are discussed in the Appendix.) These problems may be closed form "analytical" solutions of modeling problems that have to be evaluated or problems requiring iterative solutions. A *solution algorithm* is an exact prescription, defining the computational process that is applied to solve the modeling problem. An algorithm could be a mathematical recipe: a straightforward, finite sequence of logical and/or arithmetical steps, starting with given data and leading directly to the solution. Alternatively, an algorithm could be iterative and operate in cyclic fashion starting out with an initial approximation and improving the approximation in each cycle or step. That is, the solution

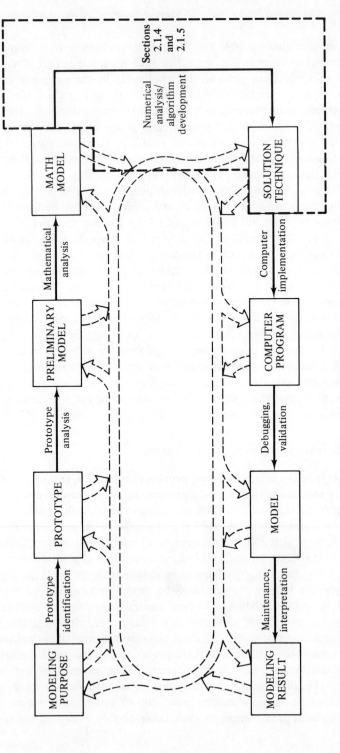

Figure 2.1.4 Guide to the discussion in Sections 2.1.4–5.

accuracy improves after each iteration. Sometimes a finite number of steps will lead to the solution, but in many cases an infinite number is (theoretically) required. All cycles consist of the same sequence of operations. Iterative algorithms require careful construction and analysis, the main elements of which are pointed out here. Specific computational algorithms are described for the three examples of Sections 2.2–2.4. Solution methods in general are beyond the scope of this presentation. For a survey of numerical methods for the solution of problems of many types, the reader is referred to Young [1973]. In this article are listed the most authoritative and current (1973) sources of computational algorithms for problems including:

1. Systems of linear and nonlinear equations,

2. Eigenvalue and eigenvector problems,

3. Approximation problems,

4. Integral equations,

5. Initial value problems and boundary value problems in ordinary and partial differential equations.

In general, an iterative algorithm then starts with an initial approximation x^0. This is the best approximation the model user has available at the outset. Also initially the iteration number k (a counter) is set at $k = 0$. A single step or cycle k of an iterative procedure operates on approximation x^k and generates approximation x^{k+1}, using information available at this stage. At the end of step k the counter is increased by one and the process is repeated. These steps generate $x^1, x^2, \ldots, x^k, \ldots, x^*$ and thereby define a motion in the space of the independent variables entering the modeling problem. The motion starts at the initial approximation x^0 and ends at the point of termination x^*. The iterative process may operate on the points $x^0, x^1, \ldots, x^k, \ldots$ directly or on related points $z^0, \ldots, z^k, \ldots$. At the typical step, say step k, the algorithm attempts to generate a successor point z^{k+1} to z^k (or x^{k+1} to x^k). There may actually be generated a whole set of points $A(z^k)$, each of which could be the successor point z^{k+1}. Each step normally involves *data preparation* and *computation*. Data from previous steps is often, but not always, used. Computation either results in the generation of (a) successor point(s) or in *termination* of the process. The iteration terminates when the candidate for successor point is identified as a solution, or if there are indications that no solution is obtainable. A solution point must have properties which are recognizable by the computational algorithm. The decision to terminate must be based on data generated by the algorithm at, or prior to, the terminating step. Otherwise, the algorithm will not be capable of terminating the computation. If x^* is a solution, the algorithm should either be incapable of generating a successor point, that is the set of successor points should be empty, or all successor points which can be generated must also be solutions. An example of a feature of a solution is that its successor points which the algorithm is capable of generating are not significantly different. That is, $\mathbf{x}^k = \mathbf{x}^*$ if $\|\mathbf{x}^{k+1} - \mathbf{x}^k\| \leqq \epsilon$ for all $\mathbf{x}^{k+1}$ in $A(\mathbf{z}^k)$ where ϵ is a small number.

In summary, the typical (kth) iterative step consists of three parts:

1. Generate and/or update the information required in (2),
2. Compute $\mathbf{x}^{k+1}$ or terminate,
3. Reset the counter $k: = k + 1$ and repeat (1).

The task of the model builder now is to define these parts. Parts (1) and (2) may consist of other algorithms involving a finite or infinite number of steps. Fortunately, solution algorithms have been developed for almost any problem that is likely to occur. The reader is referred to the aforementioned article by Young [1973] or to any of several excellent numerical analysis books for details. Whether an existing algorithm is adapted or one is built, the model builder should analyze it carefully addressing key questions as follows:

1. Does the sequence $\mathbf{x}^0, \mathbf{x}^1, \mathbf{x}^2, \ldots$ converge, and will it terminate?
2. If it does converge, how efficiently and to what limit?
3. What is the error of the (approximate) solution obtained?

Answers to these questions should provide a good basis for proper algorithm selection. If more than one algorithm can be used to solve the mathematical modeling problem, the most promising performer can be selected in the light of the answers to these questions. In Section 2.2 we outline the convergence analysis for an iterative procedure. In this case, convergence is proven by construction of two sequences of approximations, which bracket the sequence obtained by the iteration. That is, one is a sequence of lower bounds and the other is a sequence of upper bounds on the iterates. These bracketing sequences are then shown to approach the problem solution, one from below and one from above. This implies that also (the bracketed) sequence of iterates approaches the problem solution in the limit. Another method of establishing that an iteration converges is by showing that the error of solution approximation diminishes as one proceeds to compute iterates. A more theoretical treatment of this subject can be found in Chartres and Stepleman [1972].

Total computational resources required for problem solution depend on the number of iterations and the computer resources for a typical iteration. The computational cost of a single iteration can be measured in terms of the arithmetic operations and auxiliary manipulation required by the algorithm.

Our objective in general is to minimize the amount of computer resources (execution time, data transfer time, and memory storage space) used in a single step. We may, however, prefer a strategy sacrificing single-step efficiency if it will result in better performance of the overall process. Assuming that a sequence $\mathbf{x}^0, \mathbf{x}^1, \mathbf{x}^2, \ldots$ converges to a solution $\mathbf{x}^*$ we can measure the ultimate rate of convergence by the order of convergence. If there exists a real number $p \geq 1$ such that

$$\lim_{k \to \infty} \frac{\| \mathbf{x}^{k+1} - \mathbf{x}^* \|}{\| \mathbf{x}^k - \mathbf{x}^* \|^p} = c \neq 0$$

then we say that the *order of convergence* is p.

The constant c is called the *asymptotic error constant*. When $p = 1$ and $c < 1$ then the process is said to converge *linearly*. If $p = 2$ the convergence is called *quadratic*, etc. Clearly, the larger is p, the faster the sequence converges once $\| \mathbf{x}^k - \mathbf{x}^* \|$ is sufficiently small. If the initial approximation $\mathbf{x}^0$ is far from $\mathbf{x}^*$ then the initial performance of the method cannot be meaningfully measured by the order of convergence, which is a property of some iterative processes in the limit. In the initial stage of the iterative process we would like to achieve at least the following objectives:

1. Reasonably good progress of the algorithm (suitable definitions of progress are needed for different problems and methods).

2. Robust performance, i.e., no serious accumulation of inaccuracies or catastrophic breakdowns in the process of reaching the second stage where the limiting property of convergence takes over.

In looking at algorithms that terminate in a finite number of computational steps, we are most often interested in the rate of growth of the time and space required to solve problems of increasing size. By problem size is meant the number of variables, equations, constraints or other elements characterizing the quantity of data to be processed. The time needed by an algorithm to solve a problem of size n is called the *time complexity* and denoted by $0[f(n)]$. The limiting value of time complexity as the problem size increases is called the *asymptotic time complexity*. Analogously we define *space complexity*. These two parameters determine the power and efficiency of the algorithms and the size of the largest modeling problems that can be handled by currently available computers. If no other considerations are involved we should use available algorithms that offer the lowest growth rate in time and space. This is particularly important if we anticipate exercising large-scale models. Many mathematical modeling problems can be considered as follows: We are given a mathematically defined function $\mathbf{f}(\mathbf{x}) = (f_1(\mathbf{x}), f_2(\mathbf{x}), \ldots, f_m(\mathbf{x}))^T$ where $\mathbf{x} = (x_1, x_2, \ldots, x_n)^T$. The components of $\mathbf{x}$ are the data used in the model, and the components of $\mathbf{f}(\mathbf{x})$ are the model solution.

In this sense then, the purpose of the modeling exercise is to determine $f(x)$ when x changes values in some domain, or the optimal operating point, or simply to evaluate $f(x)$ for a given x.

In practice, only an approximation $\hat{x}$ to the data is known. We can also look at the computer inexact arithmetic as a factor introducing computational errors equivalent to the errors caused by solving a slightly perturbed problem, i.e., with $\hat{x}$ instead of x. In effect we have to assume that at best we can calculate $f(\hat{x})$ instead of $f(x)$ where $\hat{x}$ an approximation to x. There exist sensitive mathematical models where $f(x)$ and $f(\hat{x})$ differ greatly even for a small (in some sense) perturbation $x - \hat{x}$. Such problems are called *ill-conditioned* and no matter how carefully handled numerically they tend to give poor solutions. This ill-conditioning is not uncommon in many types of mathematical modeling problems involving, for example: linear equations, least squares problems, curve and surface fitting, roots of polynomials, boundary value problems, eigenvalue problems, differential equations and computation of recurrence relations. The extreme sensitivity of the ill-conditioned model to data perturbations may be an inherent physical property of a prototype or may be the result of the mathematical model formulation and/or the solution algorithm used. The first case is difficult to handle and requires fundamental modeling purpose revisions, or model reformulations. Not much theory related to this question has been developed. Experienced builders and users of mathematical models have claimed that ill-conditioning is in most cases introduced extraneously in modeling, e.g., via so-called "simplifying assumptions," and is usually not inherent in the physical prototype.

In the second case we should attempt to reformulate the model and the solution process. We do not expect that the ill-conditioned problem can be solved very accurately by any method, but require that the numerical algorithm should not aggravate the difficulty by producing even larger errors than those caused by ill-conditioning. Computational methods with this property are said to be *stable*. In the absence of this property, the method is *unstable*, and small input errors may result in large solution errors. In fact, an unstable method may fail even for a well-conditioned and perfectly solvable problem. A good example of a stable method is the Gauss elimination method (used to solve linear algebraic equations), providing that row and column interchanges are implemented. The same method without interchanges does not, in general, produce accurate solutions and may break down when handling a very well-conditioned problem. For additional information on the subject of ill-conditioning and numerically stable algorithms, see Fox and Mayer [1968] and Stewart [1973].

Results of mathematical modeling computations are "contaminated" by errors of several types. First are clerical errors such as misspelled formulas in computer programs, mispunched data cards, magnetic tape errors, etc.

It is useful to build input diagnostic devices into large mathematical models, particularly those requiring large amounts of data and considerable computer solutions. Examples of such devices are graphical displays of input data or computer tests designed to identify contradictory data. Other errors in modeling results are due to:

1. Model idealization and the chain of modeling approximations (e.g., discretization errors),

2. Data inaccuracies,

3. Truncation errors,

4. Computational round-off errors,

5. Numerical cancellation errors.

In most cases, it is advisable to develop an analysis of the error associated with the overall solution strategy implemented for the mathematical modeling problem. This analysis can be used in interpreting the results of the modeling experiments, that is in translating from model space to prototype space.

As pointed out in Chapter 1, models should reflect only the most essential features of the modeled problem. We introduce *model idealizations* in order to obtain a mathematically tractable formulation or because our understanding of the prototype is not complete. There is nothing wrong with such simplifications as long as the model provides useful information and fulfills the purpose for which it was built. Furthermore, the model may be reformulated once or more before it is actually used for computational experiments. Each of these reformulations may replace a problem by another whose solution approximates that of the replaced problem. These approximations can theoretically be made as accurate as desired. In reality, however, there remains an approximation error, which may not be as small as one would have liked it to be because of, say, practical limitations on available computer resources. An example of an approximation error is the error due to the discretization when a derivative is replaced by finite difference in the model.

All data generation, measurement and collection processes are error prone. It is important to be able to assess the data accuracy and accuracy of the results that these data warrant. We may, for instance, trust the trends generated by the model output but not specific numbers produced. In another case, we may require at least several-digit accuracy of the output information and should check if the available data warrant such accuracy. In analog and hybrid computation, the error problem may be reversed: we may have very accurate input data and output results which are limited to two decimal places. *Truncation errors* are introduced when we use approximate formulas with a finite number of terms instead of infinite, or terminate an iterative process before it converges to the limit (achievable only in mathematically ideal conditions). *Round-off errors* are a natural result of the finite length of computer "words." Consider as an illustration multiplication of two numbers. If the exact product of two m-digit numbers cannot be stored and used as a $2m$-digit number then a round-off error must occur. Let a and b denote two floating-point numbers and $f\ell(a \pm b)$ denote the result of floating-point addition (subtraction), then

$$f\ell(a \pm b) = (a \pm b)(1 + \epsilon)$$

where $|\epsilon| \leq 2^{-t}$
and $t =$ Number of binary digits in the mantissa of the floating-point
              representation of the number (binary computer)

This means, for example, that the approximate result of adding a and b on the computer is the exact result of adding $a(1 + \epsilon)$ to $b(1 + \epsilon)$ for some ϵ satisfying $|\epsilon| \leq 2^{-t}$. Similarly, $f\ell(ab) = ab(1 + \epsilon)$ means that the approximate result of multiplying a by b is the exact result of multiplying a by $b(1 + \epsilon)$ for some ϵ satisfying $|\epsilon| \leq 2^{-t}$ (Young [1973]). It can also be shown that

$$|f\ell(ab) - ab| \leq |ab|\epsilon$$

Since ϵ is usually between 10^{-6} and 10^{-15} there may be no reason for concern if our mathematical model does not require very extensive calculations. However, even in the case of a single multiplication, underflow or overflow may happen if ab is outside the interval of machine floating-point representation. If extensive computations are involved in modeling exercises, error accumulation and numerical instability may result. More on this subject can be found in Nickel and Ritter [1972].

Another disturbing phenomenon in numerical computation is the so-called *cancellation error* which is the result of subtraction in which the difference between operands is considerably less than the values of the operands. Suppose that we evaluate $\Delta f = f(\mathbf{x} + \Delta\mathbf{x}) - f(\mathbf{x})$ where $f(\mathbf{x})$ is a continuous function of $\mathbf{x}$. If $\Delta\mathbf{x}$ is small enough then Δf will be very inaccurate due to a loss of information carried by the several leading digits of $f(\mathbf{x})$ and $f(\mathbf{x} + \Delta\mathbf{x})$. The trailing ends of the computed values of $f(\mathbf{x})$ and $f(\mathbf{x} + \Delta\mathbf{x})$ are contaminated by round-off errors. The leading digits will cancel out. If cancellation is suspected to happen in the model, an attempt should be made to reformulate these mathematical relations where cancellation may occur.

We will close this subsection with some comments on *error estimates*. In some mathematical models, we are interested in estimating a bound for the magnitude of the error

$$\Delta f = f(\hat{\mathbf{x}}) - f(\mathbf{x}^0)$$

where $\hat{x}$ is an approximate value for x^0 and $f(x)$ is defined through a computational process and represents intermediate or final results. If Δf is large then $f(x)$ is sensitive to perturbations and great care has to be exercised in using values of $f(x)$ in the neighborhood of x^0. It can be shown that the general formula for the bound is

$$|\Delta f| \leq \sum_{i=1}^{n} \left| \frac{\partial f}{\partial x_i}(\hat{x}) \right| \cdot |\Delta x_i|$$

and the quantities $|\partial f(\hat{x})/\partial x_i|$ (or better the maximum values of $|\partial f(x)/\partial x_i|$ in the neighborhood of $\hat{x}$) can serve as a measure of the sensitivity of $f(x)$ to perturbation in the variable x. In practical computations the explicit expressions for the partial derivatives $\partial f/\partial x_i$ can be difficult to obtain and we resort to an experimental perturbation. In this test we randomly perturb the vector x and study the actual computational differences of $f(x)$. We should note here that this notion of problem sensitivity is akin to the previously defined problem ill-conditioning. In many cases it is impractical to analyze the error by tracing the floating-point arithmetic errors. However, perturbation techniques described here can almost always, be used to check the computer implementation. For more information on numerical error analysis, the reader is referred to Dahlquist and Björk [1974] and Knuth [1969].

At this stage, the model builder has constructed his solution algorithm and has determined that it possesses the right combination of the properties of convergence, efficiency, stability, and error control. In Subsections 2.2.4.5, 2.3.4.5, and 2.4.6 we discuss the solution algorithm development for the examples of Sections 2.2, 2.3, and 2.4 respectively. The model is now almost ready for computer implementation, the subject of Chapter 3. We encourage the reader to familiarize himself with mathematical software, the subject of Subsection 2.1.5, before implementing his model. By using high-quality mathematical software, which is available from a variety of sources, the model builder can improve the effectiveness and efficiency of his project.

2.1.5 Mathematical Software

Mathematical software is any documented computer program that solves a mathematical or a statistical problem. It is often possible to use mathematical software as building blocks in computer implementations of mathematical models. By incorporating available high-quality mathematical software in his computer program, the builder of a mathematical model can save time and money and also increase the overall reliability and effectiveness of his model; expensive duplication of effort can be averted by not recreating the mathematical software building block itself. Furthermore, mathematical model "debugging" and validation is much simplified and reduced in dimension. Mathematical software is already validated and model validation can focus on other elements of the model. Some of the available mathematical software has been built by expert numerical analysts and computer programmers. The quality of mathematical modeling results can be greatly enhanced by the use of optimal (see below for definition) software created by these experts.

The need for reliable and efficient standard programs handling the most common mathematical problems was recognized in the late 1950's. The necessity and usefulness of sharing software among practitioners became widely recognized and new methods of sharing software were established. In the 1960's there were more systematic attempts to develop and distribute generally useful software. Among these were manufacturer's libraries of subroutines, algorithms in journals, and codes in textbooks. Toward the end of this period specialized packages and books describing mathematical software started to appear. More attention was paid to the quality of software, documentation, user interface, and the like. Since 1970 mathematical software has emerged as a recognized field of scientific activity with its own journals, such as *ACM Transactions on Mathematical Software*, and regular meetings (e.g., Rice [1971]) devoted to this subject.

Mathematical software is currently available in libraries distributed by:

1. Hardware vendors, for example

CDC Math Science Library (developed by the Boeing Company Computing Department)

IBM Scientific Subroutine Package (SSP)

Univac Math Stat Pack.

2. Computer user groups such as

VIM, Inc., users' organization for CDC 6000 and 7000 series computers.

SHARE, IBM systems user group

USE Program Library Interchange, Univac

CUBE, Burroughs Users' Group.

3. Government laboratories and universities such as

The Numerical Analysis Group, Theoretical Physics Division, Atomic Energy Establishment, Harwell, England

The Argonne Code Center, Argonne National Laboratory, Argonne, Illinois.

4. Software vendors such as IMSL, the International Mathematical and Statistical Library (available for CDC, IBM and Univac Computers from IMSL, Houston, Texas).

5. Computer service firms make available several of the above listed mathematical software libraries and often also high-quality packages developed and maintained "in house."

Many of the algorithms embedded in the currently available mathematical software are from public-domain publications such as those of the Association for Computing Machinery (ACM), *Mathematics of Computation, Numerische Mathematik Handbook for Automatic Computation*, and many other journals. Developers of mathematical software libraries include numerical analysts and computer programmers. They concern themselves mainly with the selection and programming of algorithms, or algorithm combinations, and their incorporation in a host program which manages and controls the computation for the user. Software developers are also concerned with algorithm enhancement, testing, certification and other activities aimed at improving and assuring the quality of their library programs. For example, careful analysis of all possible outcomes of the computational process is conducted in conjunction with the actual testing to identify potential computational abnormalities. Corresponding to these critical points, error-monitoring devices are then built into the software. Mathematical software developers generally view their projects as an optimization task; the objective is to maximize a combination of speed with storage conservation. Constraints they attempt to satisfy include, according to Cody [1974], Ford and Sayers [1976], and others:

1. Capability of reliably producing accurate and valid results for all cases that are likely to arise and the ability to filter out improper cases and to react appropriately if and when they occur. This implies a requirement for technique stability, i.e., that errors introduced in the computation not grow beyond some limit.

2. Robustness, that is, the ability to detect and recover from abnormal conditions without terminating computations (Cody [1974]). In case of abnormal conditions, proper diagnostic information should be provided so as to minimize the requirement for reruns.

3. Simplicity of use, and—recently—portability from one computer to another.

Mathematical software libraries, then, consist of computer programs with the above features and appropriate user documentation. Such documentation also provides the user with guidance on optimal subroutine selection in case he has more than one option. Many libraries make available a user consultation service. New algorithms are continually developed and hence libraries must be (and many are) supported and kept up to date with respect to their dynamic environment. Needless to say that, in order to remain useful, this software must also be maintained operational on the computer hardware and system software.

Individual mathematical software programs are collected and designed with the total library structure in mind (Cody [1974]). Cate, et al. [1974] recommend a three-level mathematical software library structure. They argue that a "canned" static library will deteriorate with respect to the moving state-of-the-art, if it is not continually updated, refined, and corrected. Furthermore, if the library is used, errors may be found that should be corrected, or otherwise valuable user inputs may have to be incorporated, or new requirements may have to be accommodated. The three recommended library levels are:

1. Frequently used, standard programs in simple subroutine form. Coverage may include elementary and special function evaluation, interpolation and approximation, polynomials, nonlinear equations, quadrature and ordinary differential equations, linear algebra, probability and statistics, and nonmathematical utilities such as graphics.

2. Low-use, specialized techniques developed as needed for seldom occurring problems (e.g., "out of core" techniques for large linear algebra problems).

3. Experimental new software which has not been completely tested but may be available for limited distribution. Software will not remain permanently in this level—it will move to either level (1) or (2) or will be discarded if it is unsatisfactory.

In addition to complete libraries, several packages are available containing software in specialized areas, such as statistics, linear algebra, or nonlinear optimization.

To illustrate typical contents and capability of mathematical software libraries and packages, we provide below short descriptions of some that are in the public domain.

Harwell Subroutine Library

The Harwell Subroutine Library is maintained by the Numerical Analysis Group of the Theoretical Physics Division, Atomic Energy Research Establishment in Harwell, England. It is a library of subprograms, although some of the library routines are in fact packages of more than one subprogram. The principal language is FORTRAN but a few of the routines are written in machine code for the IBM 370/165. The library is composed mainly of mathematical and numerical routines written by members of the Harwell Numerical Analysis Group (with some exceptions). The principal function of the library is to provide Harwell computer users with good numerical facilities, but the library or individual routines can be obtained by anyone for a small handling charge. A complete list of subroutines in the Harwell Subroutine Library can be found in Hopper [1973]. This list includes subroutines in the following areas of numerical analysis:

1. Differential equations,
2. Algebraic eigenvalue and eigenvector problems,
3. Mathematical functions, random numbers and Fourier transforms,
4. Geometrical problems,
5. Integer-valued functions and integer-valued system functions,
6. Sorting and using sorted information,
7. Linear and dynamic programming,
8. Problems in linear algebra,
9. Nonlinear equation problems,
10. Input/output facilities,
11. Polynomial and rational functions,
12. Numerical integration,
13. Functions of statistics,
14. Interpolation and general approximation of functions,
15. Optimization and nonlinear data-fitting problems,
16. System facilities.

EISPACK

EISPACK is a collection of 39 FORTRAN subroutines designed to solve any standard matrix eigenvalue or eigenvector problem for real or complex matrices. It can be obtained without cost from: Argonne Code Center, Argonne National Laboratory, Argonne, Illinois. The EISPACK subroutines are modules designed to each perform one part of an eigensystem computation. More than one of these subroutines are usually required to solve most matrix eigenvalue problems. It is possible to call these routines individually. However, use of a special master routine in the EISPACK package makes this unnecessary. When this master routine is called it sets up appropriate calls to other routines required to solve a given problem. Such an ordered set of subroutine calls is called an EISPACK path. Table 2.1.2 shows the recommended paths in EISPACK (marked with ×) which identify a set of 22 basic problems that can be solved by the package. The user determines the problem he wants to solve and calls the control program EISPAC to perform the required tasks (Smith et al. [1974]).

Table 2.1.2 Recommended Paths in EISPACK

PROBLEM CLASSIFICATION / CLASS OF MATRIX	COMPLEX GENERAL	COMPLEX HERMITIAN	REAL GENERAL	REAL SYMMETRIC	REAL SYMMETRIC TRIDIAGONAL	SPECIAL REAL TRIDIAGONAL
All eigenvalues & corresponding eigenvectors	×	×	×	×	×	×
All eigenvalues	×	×	×	×	×	×
All eigenvalues & selected eigenvectors	×		×			
Some eigenvalues & selected eigenvectors		×		×	×	×
Some eigenvalues		×		×	×	×

LINPACK

This is a linear equations package being currently developed by Argonne National Laboratory in cooperation with several universities. The researchers at the various universities will produce uniform codes and provide uniform documentation. Argonne National Laboratory will collect and systematize the codes and act as the coordinator for testing the codes on various computer systems. The first effort is aimed at a collection of linear system problems which include:

1. Solving systems of linear equations when the matrix is:
   - (a) Dense and
     - i. Real general,
     - ii. Complex general,
     - iii. Real symmetric definite,
     - iv. Complex Hermitian definite,
     - v. Real symmetric indefinite,
     - vi. Complex Hermitian indefinite,
     - vii. Real and diagonally dominant,
     - viii. Complex and diagonally dominant.
   - (b) Banded and
     - i. Real general,
     - ii. Complex general,

      iii. Real symmetric definite,

      iv. Complex Hermitian definite,

      v. Real and diagonally dominant,

      vi. Complex and diagonally dominant.

  (c) Tridiagonal and

      i. Real general,

      ii. Complex general,

      iii. Real symmetric definite,

      iv. Complex Hermitian definite,

      v. Real and diagonally dominant,

      vi. Complex and diagonally dominant.

Pivoting strategies will be used where needed for numerical stability, but no scaling will be required.

    2. Computing linear least squares solutions by the QR factorization using elementary reflectors.

    3. Computing the rank of a matrix by the singular-value decomposition.

This first library should be available for use in 1978–1979. There are further problems for which good-quality software can be developed and included later in LINPACK, such as:

    1. In-core direct solution to sparse equations (large percentage of zero matrix coefficients),

    2. Out-of-core direct solution of dense equations,

    3. In-core linear programming,

    4. Iterative methods for large structured systems of equations.

MINPACK

The MINPACK project at Argonne has been initiated as a research venture to examine the organization of a collection of algorithms for nonlinear optimization. The package elements currently under development address the following problems:

    1. General nonlinear optimization: minimize an objective function $f(\mathbf{x})$ where $\mathbf{x}$ is the n-component real vector $\mathbf{x} = (x_1, \ldots, x_n)$.

    2. Nonlinear least-squares optimization: $(\min\limits_{x} \sum\limits_{i=1}^{m} f_i^2(\mathbf{x}))$.

    3. Nonlinear systems of algebraic equations.

The project has been motivated by:

    1. Increased emphasis on the analysis of data from scientific and engineering experiments and the modeling of engineering and socio-economic systems which require optimization routines.

2. A considerable gap between currently available powerful optimization algorithms and algorithms used by scientists and engineers.

Approximately twenty computer algorithms have been considered for inclusion in the first version of MINPACK. This version will be a basis for further expansion to constrained optimization problems. Another line of possible development is a poly-algorithm approach where several algorithms are used simultaneously to solve an optimization problem. The 1976 status of the product is described by Brown et al. [1975].

2.1.6 Summary

We have described in Section 2.1 the major methodological elements of mathematical model definition and analysis, including:

1. Identification of prototype, modeling purpose and statement of the modeling problem.
2. Model definition and analysis.
3. Analysis and reformulation of the mathematical modeling problem.
4. Solution algorithm development.

The process consisting of these elements combines the available information in one or more candidate models which can be implemented and used to achieve the modeling purpose. These candidate model(s) may differ in their technical features and in economics of implementation and use. The model finally selected for implementation and use should have the most favorable combination of technical and economical features. In Sections 2.2–2.4 the application of the methodology is illustrated in the development of three example models.

The candidate model(s) are ready, at this stage, for computer implementation, the subject of Chapter 3. The reader is urged again to use available mathematical software as building blocks in his computer implementation whenever appropriate.

2.2 A MATHEMATICAL MODEL OF A GENERAL TWO-DIMENSIONAL FLOW

We now consider the mathematical model of a general, steady-state, two-dimensional flow in a medium. The model could represent several different prototypes, including for example, incompressible fluid flow, water seepage, heat flow, electrical current flow, magnetic flux, flow of mechanical force or torque. The purpose of the model is to determine the refined design of the flow-conducting medium, its shape and materials. For this purpose the designer needs a detailed description of the two-dimensional flow field. Such a description can be provided by the distributed parameter model which is developed here. This development leads to a boundary value problem in a second-order partial differential equation. This problem, in turn, is approximated by a system of difference equations (linear, algebraic) for which a solution scheme is outlined. Another modeling approach, that we consider, is based on lumping of the dependent variables at selected points on a grid in the medium. This approach in fact also leads to a difference equation system, under certain circumstances to the same system as the first approach does. Both modeling methods are general and hence applicable to all the above listed flows and to many others.

2.2.1 The Prototype and the Modeling Purpose

In this subsection we identify the prototype and the phenomenon considered in this modeling study; we also define the modeling purpose.

The prototype is any flow-conducting or resisting medium. For simplicity we consider a two-dimensional case, ignoring variations in the third dimension. The medium can be thought of as very large in the third dimension, and the modeling results considered applicable per unit of distance in that direction. The medium shape is given and, again for simplicity, we assume this to be a rectangle. The medium is made of material with a given material property. The property of interest here is *conductivity*. It measures the ability of the material to conduct the flow. Alternatively, flow-conducting materials are sometimes characterized by the inverse of conductivity, i.e., *resistivity* or *dissipativity* which measure the ability to resist or to dissipate flow.

A flow takes place across the boundaries and within the medium. The flow is driven by potential differences which are measurable in the medium and on its boundaries. The potential function is a scalar function of position but not of direction. It depends on the two spatial coordinates, in this case, and does not depend on time because we are dealing with a steady flow. We assume that no other phenomena take place in the medium; specifically, that the prototype cannot store the flowing quantity, cannot act as a source, and cannot store potential. Examples of flows which fit this description and the corresponding potentials and conductivities are listed in Table 2.2.1. One of the examples of fluid flow is water seepage through a layer of permeable soil or rock. In this case one may model the outflow from the permeable layer of rock into say, a river, and the consequent drainage of the topsoil. An example of a heat flow is conduction in a furnace wall. In this case the modeler may be concerned with insulating the furnace to keep the heat inside, that is, with resistance of the medium to the heat flow.

Table 2.2.1 Examples of Flows and Corresponding Potentials and Conductivities/Resistivities

FLOW	POTENTIAL	CONDUCTIVITY/ RESISTIVITY
Incompressible fluid	Pressure/head	Viscosity
Seepage in permeable medium	Pressure/head	Permeability
Heat conduction	Temperature	Conductivity
Electrical current	Voltage	Resistivity
Magnetic flux	Magnetic potential	Reluctance
Mechanical force	Displacement/velocity	Viscous damping
Mechanical torque	Displacement/velocity	Viscous damping

The purpose of the mathematical model is to evaluate the potential function (i.e., define the flow field) at arbitrarily closely spaced points in the medium and on its boundary. The potential function may then be related to the flow and to the conductivity of the medium. Such information will enable the model user to determine whether or not his design performs properly. The design in this case consists of the medium shape and the material it is made of. If, for example, the temperature on the outside of the furnace wall is too high, another material with a lower heat conductivity may have to be used.

2.2.2 Mathematical Model Development

In this subsection we develop a mathematical model of steady flow in a two-dimensional medium. This model will provide the desired definition of the flow field.

A *distributed-parameter* model will be developed initially in order to obtain sufficient detail. Therefore a system of rectangular coordinates is used, with the two position variables x and y chosen as the two *independent variables* of the mathematical model. Time does not enter the model as an independent variable because the model is steady-state.

The fundamental law of flow in a conducting or resisting medium without storage of potential or flow states that the *flux* (flow rate) is proportional to the *potential gradient* in the direction of flow. The constant of proportionality is the material resistivity or equivalently the inverse of its conductivity. This law can be represented in the model by a dissipator element (see Subsection 2.1.2). In the x-direction for example this can be written as:

$$\frac{\partial \phi}{\partial x} = -R_x f_x \tag{2.2.1}$$

where ϕ is the potential function, R_x and f_x are the resistivity and the flux in the x-direction respectively. This flux is the flow per unit of time and per unit of area normal to the x-direction. The minus sign on the right indicates that flux is positive in the direction of decreasing potential. The fundamental law of flow has been found to apply in different sciences. In heat transfer, for example, it is referred to as Fourier's law of heat conduction (Giedt [1957]). Hydrologists refer to it as Darcy's law of water seepage in a permeable medium (Todd [1959]).

The prototype can be regarded as a system in which the flowing quantity is conserved. The model may then be derived using the shaded control area indicated in Fig. 2.2.1. The flow is broken into two component flows, one in the x-direction and the other in the y-direction. The fundamental law of flow [Eq. (2.2.1)] applies to each of these components. In this case there is no accumulation of the flowing quantity in the medium (because the medium is no reservoir of flux). Therefore the conservation of flow statement reduces to the simple balance statement that flow "in" equals flow "out": the inflows on the left and bottom sides are $\Delta y f_x$ and $\Delta x f_y$, respectively. The outflows on the right and top sides are

$$\Delta y \left(f_x + \frac{\partial f_x}{\partial x} \Delta x \right) \quad \text{and} \quad \Delta x \left(f_y + \frac{\partial f_y}{\partial y} \Delta y \right)$$

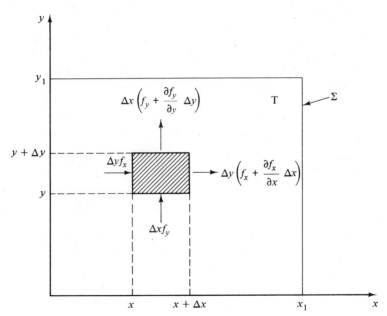

Figure 2.2.1 Diagram of flow-conducting medium, coordinate system, and control area (shaded).

respectively. Balance of flow requires that

$$\Delta y\left(f_x - f_x - \frac{\partial f_x}{\partial x}\Delta x\right) + \Delta x\left(f_y - f_y - \frac{\partial f_y}{\partial y}\Delta y\right) = 0$$

and hence:

$$\frac{\partial f_x}{\partial x} + \frac{\partial f_y}{\partial y} = 0 \qquad (2.2.2)$$

Using the law of flow (2.2.1) in (2.2.2) we obtain:

$$\frac{\partial}{\partial x}\left(\frac{1}{R_x}\frac{\partial \phi}{\partial x}\right) + \frac{\partial}{\partial y}\left(\frac{1}{R_y}\frac{\partial \phi}{\partial y}\right) = 0 \qquad (2.2.3)$$

and for the special case that the resistivity R is constant and $R = R_x = R_y$ this reduces to

$$\frac{\partial^2 \phi}{\partial x^2} + \frac{\partial^2 \phi}{\partial y^2} = 0 \qquad (2.2.4)$$

Equation (2.2.4) is a second-order partial differential equation, known as *Laplace's equation*, in two independent variables (in this case). It is one part of the mathematical model defining the potential function $\phi(x, y)$ anywhere in the region of interest T in the flow-conducting medium.

To define $\phi(x, y)$ completely and to obtain a solvable problem, two boundary conditions have to be specified for each of the independent variables. *Boundary conditions* of two types are possible, and correspondingly three types of boundary value problems can occur:

1. Boundary Value Problem of Type I (Dirichlet problem): The potential function ϕ is specified on the boundary Σ.

2. Boundary Value Problem of Type II (Neuman problem): The flow or potential gradient $\partial\phi/\partial n$ normal to the boundary is specified on the boundary Σ.

3. Mixed Boundary Value Problem: Boundary conditions of Types I and II are prescribed (i.e., $(\partial\phi/\partial n) + h(\phi - f) = 0$) on the boundary Σ. Here h and f are given functions. Notice that in the special case of $h = 0$, the mixed problem reduces to a problem of Type II.

In the simple case illustrated in Fig. 2.2.1, the boundaries are at $x = 0$, $x = x_1$, $y = 0$, and $y = y_1$, and boundary conditions of the first type specify:

$$\phi(0, y), \quad \phi(x_1, y), \quad \phi(x, 0), \quad \text{and} \quad \phi(x, y_1).$$

Boundary conditions of the second type specify:

$$\frac{\partial \phi(0, y)}{\partial x}, \quad \frac{\partial \phi(x_1, y)}{\partial x}, \quad \frac{\partial \phi(x, 0)}{\partial y}, \quad \text{and} \quad \frac{\partial \phi(x, y_1)}{\partial y}.$$

The potential function $\phi(x, y)$ constitutes a *solution* to the boundary value problem of any one of the three types, if it is a well-defined continuous function in the bounded region $T + \Sigma$, satisfying (2.2.4) inside the region T and the appropriate set of boundary conditions on the boundary Σ. A solution to the boundary value problem is also a solution to the mathematical modeling problem, because values of the potential function determine the flow everywhere in the bounded region.

To summarize: (2.2.4) together with a sufficient number of boundary conditions (two for each spatial variable) comprise the mathematical model of a general steady two-dimensional flow in a conducting medium with constant conductivity (or resistivity). The model defines a potential function and the flux components can be obtained from the subsidiary relationship (2.2.1) as a function of potential gradients. We wish to caution the reader at this stage of the development: This problem is not solvable under all circumstances, its solvability depending on exactly what combination of boundary conditions is specified. This question is further discussed in the next subsection.

2.2.3 Mathematical Modeling Problem Analysis

In this subsection we analyze the mathematical modeling problem developed for the study of steady flow in a two-dimensional medium.

A thorough analysis of the three boundary value problems involving Laplace's equation (2.2.4) is beyond the scope of this discussion. However, to illustrate the required analysis we shall consider briefly the boundary value problem of Type I. The problem is to find a potential function $\phi(x, y)$ which is defined and continuous in the bounded region $T + \Sigma$, satisfies the partial differential equation (2.2.4) inside T, and assumes prescribed boundary values on the boundary Σ.

This problem analysis attempts to establish that a unique solution to the stated problem exists and that the solution depends continuously on the boundary conditions. The importance of establishing solution uniqueness before proceeding to solve the problem is obvious: a multiplicity of solutions might defeat the purpose of mathematical modeling. Why is it necessary to establish continuous dependence on the boundary data, a property frequently also referred to as stability? Boundary conditions are often determined by measurement or by computation and may therefore involve a measurement or a computational error. It is important to assure that the problem is defined such that small errors in the boundary data result in small errors in the solution. Thereby the modeler will be assured that if he can control data errors, the modeling solution will be close to that which can be obtained under ideal conditions of no data errors.

The basis for this analysis is the property of functions $\phi(x, y)$ which are defined and continuous in a bounded region $T + \Sigma$ and satisfy Laplace's equation (2.2.4) in the interior of the region T. The property is the so-called *maximum/minimum principle*, by which $\phi(x, y)$ attains its relative maximum and minimum on the boundary of the region Σ (for proof refer to, e.g., Tikhonov and Samarskii [1963], Chapter IV). Suppose that ϕ and Φ are two continuous functions in $T + \Sigma$, satisfying Laplace's equation (2.2.4) in the interior of the region T. Then the maximum/minimum principle implies two corollaries:

1. If $\phi \leq \Phi$ on the boundary Σ then $\phi \leq \Phi$ also in the interior T. This is the case because the function $(\Phi - \phi)$ is also continuous in $T + \Sigma$ and satisfies (2.2.4) in T and $\Phi - \phi \geq 0$ on the boundary Σ (given condition). Therefore by the maximum/minimum principle $\Phi - \phi \geq 0$ also in the interior T, which proves this corollary.

2. If $|\phi| \leq \Phi$ on the boundary Σ then $|\phi| \leq \Phi$ also in the interior T. This can be verified by application of Corollary (1) to $- \Phi$, ϕ and Φ satisfying $-\Phi \leq \phi \leq \Phi$ on the boundary Σ.

We return now to the analysis of the boundary value problem of Type I. Solution *existence* is implied when a solution can be constructed. This is done below. To establish *uniqueness* assume two distinct solutions ϕ_1, and ϕ_2. By definition these potential functions are continuous in $T + \Sigma$, satisfy Laplace's equation in T and the boundary conditions of the first type on Σ. Form a difference function

$$\Phi = \phi_1 - \phi_2$$

The function Φ thus formed satisfies Laplace's equation in the interior T, is continuous in the bounded region $T + \Sigma$ and is identically zero on the boundary Σ. The latter is true because ϕ_1, and ϕ_2 each satisfy the same conditions on the boundary. That is,

$$\Phi = \phi_1 - \phi_2 \equiv 0 \quad \text{on } \Sigma.$$

By the maximum/minimum principle, however, Φ must attain its maximum and minimum on the boundary. Therefore $\Phi \equiv 0$ also in the interior T, which implies $\phi_1 \equiv \phi_2$ thereby establishing uniqueness.

We proceed next to prove that the boundary value problem of Type I is *stable*, i.e., its solution depends continuously on the boundary data.

Assume again two distinct solutions ϕ_1, and ϕ_2. By definition these potential functions are continuous in $T + \Sigma$, satisfy Laplace's equation in T and the boundary conditions of the first type on Σ. Suppose that $|\phi_1 - \phi_2| \leq \epsilon$ on the boundary Σ, where ϵ is an arbitrarily small number. To establish stability of this boundary value problem we have to show that $|\phi_1 - \phi_2| \leq \epsilon$ also in the interior T. To see that this inequality is satisfied we use Corollary (2) of the maximum/minimum principle setting $\Phi \equiv \epsilon$ (this is possible because ϵ is also a solution to the boundary value problem). Thereby $|\phi_1 - \phi_2| \leq \Phi$ on the boundary Σ implies that $|\phi_1 - \phi_2| \leq \Phi$ also in the interior T, proving stability.

This completes the analysis of boundary value problem of Type I which was presented as an illustration of the type of analysis required for these problems. After the successful conclusion of such an analysis, our modeling problem is ready for solution.

2.2.4 Finite-difference Solution Process Development

This subsection deals with the development of a finite-difference solution process for the modeling problem of steady flow in a two-dimensional medium.

A solution for any of the three boundary value problems defined above can be obtained by a finite-difference method. In our discussion of such a method we follow the presentation given by Collatz [1960], Chapter V.

First a rectangular mesh is introduced into the region of interest with lines defined as

$$\left.\begin{array}{l} x_i = x_0 + ih \\ y_j = y_0 + jk \end{array}\right\} \quad i, j = 0, \pm 1, \pm 2, \ldots \qquad (2.2.5)$$

where h, k, x_0, and y_0 are given (e.g., $x_0 = y_0 = 0$). The values of the unknown potential function at the nodal points of this mesh are denoted by

$$\phi_{ij} = \phi(x_i, y_j)$$

and approximations of these function values are denoted by

$$\Phi_{ij} \cong \phi_{ij} \qquad (2.2.6)$$

Partial derivatives may be approximated by using differences between function values of the mesh points:

$$\frac{\partial \phi_{ij}}{\partial x} \cong \frac{\Phi_{i+1,j} - \Phi_{i-1,j}}{2h}$$

$$\frac{\partial^2 \phi_{ij}}{\partial x^2} \cong \frac{\Phi_{i+1,j} - 2\Phi_{ij} + \Phi_{i-1,j}}{h^2}$$

and similarly for $\partial\phi/\partial y$ and $\partial^2\phi/\partial y^2$. Then the finite-difference approximation of Eq. (2.2.4) is:

$$\frac{1}{h^2}(\Phi_{i+1,j} - 2\Phi_{ij} + \Phi_{i-1,j}) + \frac{1}{k^2}(\Phi_{i,j+1} - 2\Phi_{ij} + \Phi_{i,j-1}) = 0 \quad (2.2.7)$$

This approximation is valid for interior points of the region T whose four neighbors are either interior points or boundary points (on Σ).

Finite-difference approximations for the boundary points can be developed in a similar manner. A different formula must be used for points in the region and on the boundary $T + \Sigma$, for which not all four neighbor points are also in the region and on the boundary. One example of such an approximation formula is linear interpolation along the mesh line (for details see Collatz [1960]). On mesh points coinciding with boundary points the same value is used for Φ as is prescribed by the boundary conditions of Type I or III. In the special case of a square mesh, i.e. $h = k$, approximation (2.2.7) reduces to

$$4\Phi_{ij} = \Phi_{i-1,j} + \Phi_{i+1,j} + \Phi_{i,j-1} + \Phi_{i,j+1} \quad (2.2.8)$$

This equation is often referred to as *Laplace's difference equation*. It simply states that Φ_{ij} equals the average of the values of Φ at the four neighboring mesh points.

Before proceeding to develop a solution scheme, we must examine whether a unique solution to the finite-difference approximation exists. In general, the existence of a unique solution to an approximation of a mathematical model is not guaranteed by the existence of a unique solution to the original problem. It can be shown (Collatz [1960]) that there is no difficulty with the case of interest in this respect: the system of equations constituting finite difference approximations for the boundary value problems of interest always possesses a unique solution. Furthermore, Garabedian [1964] shows that as the mesh size h is refined, the error of approximation

$$\epsilon_{ij} = \phi_{ij} - \Phi_{ij}$$

approaches zero like h^2. In addition to the approximation error we have to contend also with a computational rounding error, because only a finite number, say ℓ, of decimal places are retained. It is further shown then that Φ_{ij} as computed on a mesh of size h while retaining ℓ decimal places converges to the exact solution ϕ if the mesh size approaches zero while at the same time ℓ goes to infinity.

The region of interest T may be large and a fine mesh may be required to provide the solution detail necessary for the modeling purpose. Therefore there will be many mesh points, giving rise to a large system of difference equations. One would normally use iterative methods to solve such a system.

An iterative method starts with an initial approximation Φ_{ij}^0, $i, j = 0, 1, \ldots$, and determines successive approximations $\Phi_{ij}^1, \Phi_{ij}^2, \ldots, i, j = 0, 1, \ldots$. The initial approximation and all subsequent approximations at boundary points coincide with the specified boundary values wherever these are specified.

For the case of interest, Eq. (2.2.8), successive approximations may be obtained for example from:

$$\Phi_{ij}^{u+1} = \tfrac{1}{4}(\Phi_{i+1,j}^u + \Phi_{i,j+1}^u + \Phi_{i-1,j}^u + \Phi_{i,j-1}^u) \qquad (2.2.9)$$

Next the following three questions need to be explored:

1. Does this iterative procedure converge, that is: do Φ_{ij}^u computed via (2.2.9) converge toward Φ_{ij} defined by (2.2.8) as u approaches infinity?

2. What is the convergence efficiency and how can it be improved?

3. What is the error of the approximate solution?

Collatz [1960] proves convergence of the procedure and derives an estimate of the solution error. The iterative procedure is stopped and considered converged to a solution when the successive approximations no longer change significantly. Garabedian [1964] also examines these questions. He denotes by α and β the lower and upper bounds on Φ_{ij} and assumes that the initial approximation is chosen to satisfy

$$\alpha \leq \Phi_{ij}^0 \leq \beta$$

Suppose that we start with α, Φ_{ij}^0 and β and construct three sequences of successive approximations to Φ_{ij} denoted by $\underline{\Phi}_{ij}^u$, Φ_{ij}^u, and $\bar{\bar{\Phi}}_{ij}^u$, respectively, where $u = 1, 2, \ldots$. We have

$$\underline{\Phi}_{ij}^0 \leq \Phi_{ij}^0 \leq \bar{\bar{\Phi}}_{ij}^0$$

If (2.2.9) is used to construct these sequences, the successive approximations are simply the average of the previous approximations at the neighboring points and therefore

$$\underline{\Phi}_{ij}^u \leq \Phi_{ij}^u \leq \bar{\bar{\Phi}}_{ij}^u$$

Garabedian [1964] shows that as u increases toward infinity $\underline{\Phi}_{ij}^u$ always increases or remains unchanged, $\bar{\bar{\Phi}}_{ij}^u$ always decreases or remains unchanged while both converge toward Φ_{ij} in the limit and therefore also

$$\lim_{u \to \infty} \Phi_{ij}^u = \Phi_{ij} \qquad (2.2.10)$$

Next we turn to the question of computational efficiency, which depends among other things on the specific sequence of points at which (2.2.9) is applied. The sequence of iterations implied by

$$\Phi_{ij}^{u+1} = \tfrac{1}{4}(\Phi_{i-1,j}^{u+1} + \Phi_{i,j+1}^{u+1} + \Phi_{i+1,j}^{u} + \Phi_{i,j-1}^{u}) \qquad (2.2.11)$$

and referred to as the *Gauss-Seidel method* converges about twice as fast as the sequence implied by (2.2.9). This formula implies that the sequence of iterations proceeds from left to right and then from top to bottom in the region. That is, starting with the highest j, points are traversed in the order $i = 0, 1, \ldots$ and when all values of i for that j are exhausted, the process is repeated with $j = j - 1$ and so on until all values of j are exhausted. At that point the iteration cycle starts all over. The situation at the typical stage (i, j) is:

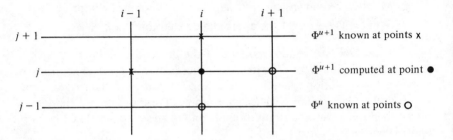

Even further acceleration of convergence can be obtained if

$$\Phi_{ij}^{u+1} = \Phi_{ij}^{u} + \frac{w}{4}(\Phi_{i-1,j}^{u+1} + \Phi_{i,j+1}^{u+1} + \Phi_{i+1,j}^{u} + \Phi_{i,j-1}^{u} - 4\Phi_{ij}^{u}) \quad (2.2.12)$$

which is referred to as *successive over relaxation*, with $w > 1$ being the *relaxation factor*. Various computer implementations of (2.2.12) exist, as does a vast literature on the proper choice of the relaxation factor. For detailed treatment of these methods the reader is referred to Varga [1962], Householder [1964], and Young [1971].

A different strategy is employed by the so-called elimination (or direct) methods for the solution of linear equation systems such as dealt with here, i.e., equations like Laplace's difference equation (2.2.8) for all values of i and j. Elimination methods such as Gauss, Choleski, and others are based on factorization of A as the product LU where A is the coefficient matrix of the linear system $A\mathbf{x} = \mathbf{b}$, L is lower triangular and U is upper triangular. Then the system $LU\mathbf{x} = \mathbf{b}$ is easily solved in the two steps:

1. $L\mathbf{y} = \mathbf{b}$
2. $U\mathbf{x} = \mathbf{y}$

There are several advantages in using direct methods:

1. In practice Gauss elimination with row and column interchanges is a numerically stable and accurate algorithm.

2. If A is not too ill-conditioned we can use iterative refinement to solve $A\mathbf{x} = \mathbf{b}$ with the maximum accuracy available on a given computer.

3. They enable us to take advantage of special problem features like sparsity of the coefficient matrix A. For developments in this area see Willoughby [1968], Reid [1970], and Rose and Willoughby [1972].

4. Computational speed, provided that sufficient primary computer memory is available.

The Choleski variant is used when A is symmetric and positive definite. In this case no row interchanges are needed and we get $A = LL^T$.

A numerically safe method to solve symmetric but indefinite systems has been given by Bunch and Parlett [1971]. The use of direct methods for solving partial differential equations by difference of finite-element methods has been described by Widlund [1972], Martin and Rose [1972], and George [1972].

This concludes the development of the finite-difference solution process for the modeling problem of steady flow in a two-dimensional medium. This model is now ready for computer implementation.

2.2.5 Finite-element Solution Process Development

This subsection deals with an alternative modeling approach for the steady flow in a two-dimensional medium. We will outline the finite-element method, an approach leading to the same difference equations that were previously obtained for our problem. In this case the difference equations arise from approximations of the function satisfying the differential equation. This approximation is made at selected points in the medium of interest at which the values of dependent variables are lumped. The finite-element method was developed for the solution of modeling problems arising in structural mechanics. It is discussed here because highly efficient computer implementations of the method are available on practically every kind of large-scale digital computer. These codes are suitable for very large problems, involving many mesh points (or node points as the points are referred to in this context). Several of these codes have convenient data handling features and incorporate highly efficient computational techniques. The method can also handle:

1. Related and more complicated problems than Eq. (2.2.4) such as Eq. (2.2.3).

2. Boundary conditions of types other than I, i.e., II or mixed.

3. Irregularly shaped boundaries.

4. Higher-order approximations of the potential function within each element in terms of the nodal values.

We consider here the boundary value problem of Type I involving Laplace's equation (2.2.4) in two dimensions. The method is applicable also to the other two types of boundary value problems, but the discussion of these is slightly more complicated. Our departure point is Euler's theorem of variational calculus (see, e.g., Zienkiewicz [1971], Appendix 6). Using this theorem it can be shown that the solution of our boundary value problem minimizes the integral

$$I(\phi) = \iint\limits_{T+\Sigma} \frac{1}{2R}\left[\left(\frac{\partial\phi}{\partial x}\right)^2 + \left(\frac{\partial\phi}{\partial y}\right)^2\right] dx\, dy \qquad (2.2.13)$$

on the set of functions which take the same boundary value $f(x, y)$ on Σ. R is the resistivity of the medium as before. This means that the problem of minimizing $I(\phi)$ subject to boundary conditions of Type I is equivalent to our boundary value problem of Type I. The finite-element method can be used to obtain an approximate solution to this problem. For this purpose the flow-conducting medium is divided by a grid into *finite elements*. These elements are assumed to be interconnected at given points on the element boundaries called *node points*. The finite-element method finds approximate values of the potential function ϕ at these node points. The potential function within each element can then be evaluated using a given relationship which specifies the function in the element in terms of its values at the element's nodes. (This is similar to the way in which the potential function can be evaluated at a point which is not on the mesh of a finite-difference solution scheme.) The next and final step is then as before to compute the flux everywhere in the region. This can of course be accomplished using (2.2.1).

Suppose then that we have node points $i, j, k, \ldots$ and define the approximation Φ of the unknown potential function ϕ as:

$$\phi \cong \Phi = [N_i, N_j, \ldots][\Phi_i, \Phi_j, \ldots]^T = [N][\Phi^e]^T \qquad (2.2.14)$$

where $\Phi_i, \Phi_j, \ldots$ are approximations of ϕ on the node points $i, j, \ldots$ respectively and where $N_i, N_j, \ldots$ are matrices whose elements are in general functions of position. The function Φ assumes the prescribed boundary values on these nodes which lie on the boundary Σ. Using the approximation Φ in the integral (2.2.13) and denoting this approximation by I^e we can find the minimum from a set of equations such as:

$$\frac{\partial I^e}{\partial \Phi_i} = \iint\limits_{\text{element}} \frac{1}{R}\left[\frac{\partial\Phi}{\partial x}\frac{\partial}{\partial\Phi_i}\left(\frac{\partial\Phi}{\partial x}\right) + \frac{\partial\Phi}{\partial y}\frac{\partial}{\partial\Phi_i}\left(\frac{\partial\Phi}{\partial y}\right)\right] dx\, dy = 0$$

and so on for j, k, etc. But from (2.2.14)

$$\frac{\partial \Phi}{\partial x} = \left[\frac{\partial N_i}{\partial x}, \frac{\partial N_j}{\partial x}, \ldots \right] [\Phi^e]^T \quad \text{and} \quad \frac{\partial}{\partial \Phi_i}\left(\frac{\partial \Phi}{\partial x} \right) = \frac{\partial N_i}{\partial x}$$

and so on for j, k, etc. Thus for one element we have the equations minimizing (2.2.13)

$$\frac{\partial I^e}{\partial \Phi^e} = \left[\frac{\partial I^e}{\partial \Phi_i}, \frac{\partial I^e}{\partial \Phi_j}, \ldots \right]^T = [h^e][\Phi^e]^T = 0 \qquad (2.2.15)$$

where the matrix $[h^e]$ is referred to as the *stiffness matrix* whose elements are given by

$$h_{ij}^e = \iint\limits_{T^e + \Sigma^e} \frac{1}{R}\left[\frac{\partial N_i}{\partial x}\frac{\partial N_j}{\partial x} + \frac{\partial N_i}{\partial y}\frac{\partial N_j}{\partial y} \right] dx\, dy$$

Taking into account all elements (i.e., all relationships (2.2.15)), the set of approximating equations is given by:

$$\frac{\partial I}{\partial \Phi} = \sum_{\text{elements}} \frac{\partial I^e}{\partial \Phi^e} = [H][\Phi]^T = 0 \qquad (2.2.16)$$

where

$$H_{ij} = \sum h_{ij}^e$$

with the summation going over all elements. The fluxes defined by (2.3.1) can be obtained from

$$\left[\frac{\partial \Phi}{\partial x}, \frac{\partial \Phi}{\partial y} \right]^T = [S^e][\Phi^e]^T$$

where for the ith node

$$S_i^e = \left[\frac{\partial N_i}{\partial x}, \frac{\partial N_i}{\partial y} \right]^T$$

To make the analysis more definite and to obtain specific expressions for the matrices N_i, N_j, ... we consider the special case of triangular elements illustrated in Fig. 2.2.2. For this case (2.2.14) becomes

$$\Phi = [N_i, N_j, N_m][\Phi_i, \Phi_j, \Phi_m]^T = [N][\Phi^e]^T$$

and $\quad N_i = (a_i + b_i x + c_i y)/2\Delta$
$\quad a_i = x_j y_m - x_m y_j$
$\quad b_i = y_j - y_m$
$\quad c_i = x_m - x_j$

$$2\Delta = 2 \iint\limits_{\text{element}} dx\, dy = \det \begin{vmatrix} 1 & x_i & y_i \\ 1 & x_j & y_j \\ 1 & x_m & y_m \end{vmatrix}$$

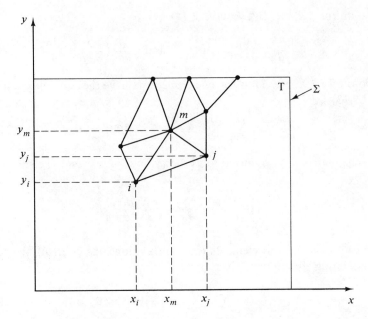

Figure 2.2.2 Division of a two-dimensional region into triangular elements.

that is, Δ is the area of the triangular element ijm. In this case the contribution of a node to the set of minimizing equations is

$$\frac{\partial I^e}{\partial \Phi_i} = \frac{1}{(2\Delta)^2 R} \iint\limits_{\text{element}} ([b_i, b_j, b_m][\Phi^e]^T b_i + [C_i, C_j, C_m][\Phi^e]^T C_i) \, dx \, dy$$

and so on. Any element contributes only to the three partial differentials associated with its nodes so that for one triangular element we have the equations minimizing (2.2.13)

$$\frac{\partial I^e}{\partial \Phi^e} = \left[\frac{\partial I^e}{\partial \Phi_i}, \frac{\partial I^e}{\partial \Phi_j}, \frac{\partial I^e}{\partial \Phi_m}\right]^T = [h^e][\Phi^e]^T = 0$$

and the stiffness matrices $[h^e]$ are given by

$$[h^e] = \frac{1}{4 \, \Delta R} \begin{bmatrix} b_i b_i & b_i b_j & b_i b_m \\ b_j b_i & b_j b_j & b_j b_m \\ b_m b_i & b_m b_j & b_m b_m \end{bmatrix} + \frac{1}{4 \, \Delta R} \begin{bmatrix} c_i c_i & c_i c_j & c_i c_m \\ c_j c_i & c_j c_j & c_j c_m \\ c_m c_i & c_m c_j & c_m c_m \end{bmatrix}$$

Notice that the two component matrices are symmetric and that the stiffness matrix depends only on R and on the geometry. Taking into account the contribution from all elements we obtain

$$\frac{\partial I}{\partial \Phi} = \sum_{\text{elements}} \frac{\partial I^e}{\partial \Phi^e} = \sum_{\text{elements}} \sum_{\text{nodes}} h_{im}\Phi_m = 0 \qquad (2.2.17)$$

and

$$\left[\frac{\partial \Phi}{\partial x}, \frac{\partial \Phi}{\partial y}\right]^T = \begin{bmatrix} b_i & b_j & b_m \\ c_i & c_j & c_m \end{bmatrix}[\Phi_i, \Phi_j, \Phi_m]^T$$

It can be shown that the detailed equations (2.2.17) are identical, for certain "regular" elements, with the equations of the finite-difference scheme for the same mesh points. These "regular" elements are illustrated in Fig. 2.2.3. Because the equations are the same, our foregoing discussion of solution uniqueness, of the diminishing error of approximation as the mesh size is refined, and of solution methods does apply here also.

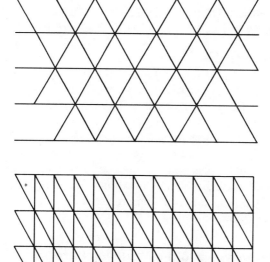

Figure 2.2.3 Finite-element schemes for which the set of approximation equations is identical with the finite-difference approximation equations for the same mesh points.

2.2.6 Summary

A mathematical model of a general steady two-dimensional flow was developed in this section. Its purpose is to obtain a detailed description of the flow field. First the prototype and the modeling purpose were defined and then a mathematical model was developed. The model is a distributed-parameter model. It gives rise to three kinds of boundary value problems involving Laplace's equation, a second-order partial differential equation. The existence of a unique solution to one problem type and the continuous dependence of this solution on the boundary conditions was discussed. Then two approximate solution methods were developed. One method is based on a finite-difference approximation and the other on a finite-element discretization (i.e., basically a lumped model). Both lead, under certain circumstances, to the same system of linear algebraic equations. The convergence of these approximations to the solution of the original boundary value problem was demonstrated. Then some solution schemes were outlined and their convergence and efficiency discussed.

2.3 A MATHEMATICAL MODEL OF STEADY NONLINEAR FLOW IN A PIPE NETWORK

In this section we develop a mathematical model of a steady flow of fluid in a pipe network. The purpose of the network is to transmit fluid from a set of sources (one or more) to a set of consumers (one or more). The purpose of the model is to assist in the pipe network design. The model relates the network structure, element characteristics, flows, and pressures in the network. The modeling problem is a set of nonlinear algebraic and transcendental equations. The solution method is in two parts: one part generates a so-called computational model and the other consists of an iterative method for solving the set of nonlinear model equations. The mathematical model and the solution of the modeling problem are general and can be used for many different network flow problems.

2.3.1 The Prototype and the Modeling Purpose

In this subsection we identify the prototype and the phenomenon considered in this modeling study; we also define the modeling purpose.

The prototype is a network of interconnected pipes, fittings, pumps and other elements. Interconnections of network elements occur at *nodes*. The manner in which the elements are interconnected is referred to as the network *structure* or *layout*. The type of each network element is known (e.g., pipe, fitting, pump, etc.) and also its shape, length, and characteristic size (e.g., pipe diameter) are known. The network is used to transmit fluid from a set of sources, called inflow nodes, to a set of consumers, called outflow nodes. There may be one or more inflow nodes and one or more outflow nodes. Examples of inflow nodes include storage tanks, reservoirs and pumps. Outflow nodes could be fire hydrants or households for example. The fluid in the network may be water, fuel, oil or any other fluid that can be modeled as an incompressible fluid. It flows through the network elements due to differences in pressure at the nodes and is characterized by a specific weight γ and a dynamic viscosity μ.

The purpose of the mathematical model is to assist the designer in the network layout and in selection of network elements and their sizes, that is, sizes of pipes, pumps, and other elements and the manner in which they are interconnected. The model must relate the network structure, characteristic sizes of elements, flows and pressure drops (or head losses). It must be capable of simulating the operation of designed networks to enable the designer to predict their performance. The network performance simulation problem is that of determining pressures and flows throughout the network statisfying the mathematical model relationships under specified operating conditions. The specified conditions could, for example, include outflows at certain nodes and pressures at certain nodes. The model user-engineer may wish to modify his design several times and predict the performance of all these prototype variations. Therefore the model must be flexible and readily applicable to a variety of network designs and operating conditions. Other, more standard requirements are that the model must be reliable, easy to use, and efficient in its use of computer resources. It is also essential that accuracy of modeling results be commensurate with data accuracy. Note that only network interconnection, flows, pressures, etc., were identified as relevant to the modeling objectives; structural strength properties of the steel pipes, for example, are of no interest. It should be noted that many other engineering, economical and operational analysis prototype systems can be modeled as nonlinear or (as a special case) linear networks. Examples include:

1. Electrical networks in which current flows through elements such as resistances.

2. Systems in which heat is transferred from one element to another.

3. Mechanical or structural networks in which forces cause displacements.

4. Commodity flow or transportation networks in which the driving "forces" are price differentials.

5. PERT network models of project status in which elements represent activities and nodes represent events. In this case, the node potential represents the time at which the event, represented by the node, takes place.

2.3.2 Mathematical Model Development

In this subsection we develop a mathematical model of steady nonlinear flow in a pipe network. This model will provide the desired simulation of the network performance.

The following discussion is based on outlines of articles by Jacoby and Twigg [1968] and Gagnon and Jacoby [1975]. The problem statement and the mathematical model are applicable to networks for any incompressible flow. The modeling method is applicable to any nonlinear network model. The model in this case is a *lumped-parameter* model in which no spatial co-ordinates enter, and all variables are represented by their values at network nodes or through network elements. A steady or quasi-steady state case of network flow is considered here and hence all time dependencies are ignored in the model. The mathematical model of the network consists of a set of equations which relate the pressures and flows throughout the network. These equations include a relationship between headloss, flow, and the element characteristic parameters for each element, and an equation of conservation of flow for each node. In addition, the model expresses the single-valuedness of pressure at any node.

Suppose that the network has ℓ elements denoted by index i, ($i = 1$, $2, \ldots, \ell$) and m nodes denoted by index j ($j = 1, 2, \ldots, m$). Each element is arbitrarily assigned an orientation (i.e., an initial or input node and a terminal or output node). The quantity q_i denotes the flow through element i and the convention $q_i > 0$ is used to imply flow from the input node to the output node and conversely $q_i < 0$ denotes flow from output to input node. The quantity Q_j denotes the external flow at node j and the convention $Q_j > 0$ is used to imply an inflow and $Q_j < 0$ implies an outflow. H_j or P_j denote hydraulic head or pressure respectively at node j.

Four types of problems arise, corresponding to four types of *boundary conditions*. These conditions may be specified at k *boundary nodes*, where k satisfies $1 \leq k \leq m$. (If $k < 1$ the problem does not make sense and obviously $k \gg m$, i.e., the number of boundary nodes cannot exceed m the total number of network nodes.) Possible boundary value problems and the corresponding boundary conditions are (see Birkhoff and Diaz [1956] and compare Subsection 2.2.2):

1. Boundary Value Problem of Type I (Dirichlet problem): Boundary conditions of Type I, namely pressures $\bar{P}_j$ or hydraulic heads $\bar{H}_j$ are prescribed at boundary nodes.

2. Boundary Value Problem of Type II (Neuman problem): Boundary conditions of Type II, namely external node flows $\bar{Q}_j$ are prescribed at boundary nodes.

3. Boundary Value Problem of Type I′: Pressures or hydraulic heads are prescribed at boundary nodes as functions of flow, i.e., $\bar{P}_j(Q_j)$ or $\bar{H}_j(Q_j)$. Notice that this is a general problem of which the problem of Type I is a special case.

4. Boundary Value Problem of Type II′: Flows at boundary nodes are prescribed as functions of pressures or hydraulic heads, i.e., $\bar{Q}_j(P_j)$ or $\bar{Q}_j(H_j)$.

Notice that this is a general problem of which the problem of Type II is a special case.

5. Mixed Boundary Value Problem: Boundary conditions of Types I and II and/or I' and II' are prescribed at boundary nodes.

Most problems of practical interest are of the mixed type: Reservoirs and storage tanks are modeled as conditions of Type I. Outflows at points of fluid consumption are modeled as conditions of Type II. Pumps can be modeled as conditions of Type I' or Type II'.

Conservation of flow at each node (the so-called first Kirchhoff law or "node law") can be expressed by:

$$\sum_{i=1}^{\ell} \delta_{ij} q_i + Q_j = 0 \qquad j = 1, 2, \ldots, m \qquad (2.3.1)$$

where δ_{ij} is zero if the ith element is not connected to the jth node, otherwise δ_{ij} is $+1$ if the ith element is oriented toward the jth node, and -1 if it is oriented away from the jth node. The quantity Q_j is the flow at the jth node, i.e.,

$$Q_j = \begin{cases} 0 \text{ at an interior node} \\ \bar{Q}_j, \text{ the specified flow at a boundary node.} \end{cases}$$

The node flow $\bar{Q}_j$ may be specified as a constant (boundary condition of Type II) or as a function of node pressure $\bar{Q}_j(P_j)$ or hydraulic head $\bar{Q}_j(H_j)$ (boundary condition of Type II').

The equation setting the sum of all external flows equal to zero

$$\sum_{j=1}^{m} Q_j = 0 \qquad (2.3.2)$$

completes the statement of conservation of flow. This corresponds to a statement of flow conservation in the entire network, implying for example that leaks are not accounted for in the model.

Conservation of flow in a network can also be modeled in terms of flows in a set of independent network loops. To develop such a model, "pseudo-loops" connecting inflow and outflow nodes must be added to the network model. This approach which would lead to a dual formulation of the problem is not used here and therefore not described in detail.

Each of the ℓ network elements is represented in the model by a characteristic relationship. This is a relationship between flow q_i through the element (the through variable) and pressure drop Δp_i across the element (the across variable). Two equivalent forms are possible:

$$\Delta p_i = F_i(q_i, \alpha_i) \qquad i = 1, 2, \ldots, \ell \qquad (2.3.3)$$

or

$$q_i = G_i(\Delta p_i, \beta_i) \qquad i = 1, 2, \ldots, \ell \tag{2.3.3'}$$

F_i and G_i are known *element characteristic functions* and α_i and β_i are parameters whose values must be specified to define the functions. These parameters are usually empirical and depend upon the properties of the fluid and lengths, diameters and loss coefficients for pipes, for example. The functions (2.3.3), (2.3.3′) may be specified in analytic form, or as a table of values. If the functions are given in tabular form, a rule for interpolating between the tabulated values must also be given. In general, the model for incompressible fluid flow in network elements is a dissipator (see Subsection 2.1.2) of the form

$$F(q_i, \alpha_i) = \alpha_i q_i^a \tag{2.3.4}$$

where a is a constant. For water flow through pipe elements we may use the empirical Hazen-Williams formula which can be stated to account for the flow direction convention:

$$F(q_i, \alpha_i) = \alpha_i \, (\text{sign } q_i) \, |q_i|^{1.85} \tag{2.3.4'}$$

where α_i depends on the units, on the pipe diameter, length, and on a "resistance coefficient." The latter is determined empirically. Details can be found in King [1954], pp. 6–11.

The mathematical model developed here can also handle network elements other than pipes. In fact any element which can be characterized by an expression of the form given by (2.3.3) or (2.3.3′) is admissible. It should be emphasized, however, that a unique solution to the mathematical modeling problem might not exist if pressure differences are not continuous and strictly increasing functions of flow. Any continuous, strictly increasing function F_i as defined by (2.3.3) has a unique, continuous, and strictly increasing inverse function G_i as defined by (2.3.3′). Therefore an equivalent necessary condition for existence of a unique solution to our problem is that flows be continuous and strictly increasing functions of pressure differences. This subject is further discussed below in the subsection dealing with mathematical model analysis. It should also be noted that (2.3.3) or (2.3.3′) must be so specified that a function value can be obtained for any value of the argument that may arise. An iterative method (see below) is used to solve the nonlinear modeling problem. Flows or headlosses encountered in this iterative process may not lie in physically meaningful ranges of (2.3.3) or (2.3.3′), even if they do so at the solution. Consequently, it is important that these relationships be artificially defined beyond their physically meaningful ranges in such a way as to preserve continuity and the property of strictly increasing flow as a function of pressure difference. Failure to provide these definitions may cause the iterative procedure to fail to converge to a solution. It should be noted that the Hazen-Williams formula (2.3.4') satisfies these conditions.

The single-valuedness of pressure at any network node is assured by assigning to each node one variable P_j representing the pressure at the node. The so-called second Kirchhoff law or "loop law" can be modeled by substituting for Δp_i in (2.3.3) or (2.3.3') the expression

$$\Delta p_i = (P_{\text{in}_i} - P_{\text{out}_i}) = -\sum_{j=1}^{m} \delta_{ij} P_j, \quad i = 1, \ldots, \ell \qquad (2.3.5)$$

where P_{in_i} and P_{out_i} are the pressures at the input and output nodes of element i respectively and δ_{ij} is as previously defined. Either of these nodes, or both nodes may be boundary nodes at which the pressure P_j is specified as a boundary condition $\bar{P}_j$ or $\bar{P}_j(Q_j)$. An alternative way of assuring the uniqueness of pressure at any node is to require that the sum of the pressure differences Δp_i around any loop be zero. This model is used in connection with fictitious loop flows and requires special consideration for elements which do not belong to loops (e.g., by adding pseudoloops).

Equations (2.3.1), (2.3.2), and (2.3.3) with (2.3.5) used for Δp_i in (2.3.3) and a set of appropriate boundary conditions comprise the *mathematical model* of steady flow in a pipe network. The model relates the network structure, its element characteristics, the flows, and the pressure drops throughout the network. A *solution* to the mathematical modeling problem consists of a set of flows q_i and pressure drops $\Delta p_i, i = 1, \ldots, \ell$ and node pressures P_j and flows $Q_j, j = 1, \ldots, m$ satisfying the mathematical model relationships and boundary conditions.

2.3.3 Mathematical Modeling Problem Analysis

In this subsection we consider the existence of a unique solution to the above stated mathematical modeling problem. The existence of a unique solution to this problem depends in part on the specified boundary conditions. There may be no solution if too many conditions are specified, and infinitely many if too few are specified. Solution uniqueness also depends

1. On the relationship between the flow through and the pressure difference across the network elements, and

2. On the relationship between flow and pressure for boundary conditions of Type I' or II' whenever these are specified.

It is easy to show that (2.3.1) and (2.3.2) contain redundant information. Since each q_i appears in two of the equations (2.3.1)—once with a positive

δ_{ij} and once with a negative δ_{ij}—it follows that the sum of the m equations (2.3.1) reduces to Eq. (2.3.2). Indeed the negative of any one of the equations can be obtained from the sum of the remaining equations provided we multiply Eq. (2.3.2) by -1. As a consequence, flows which satisfy m of the $(m+1)$ Eqs. (2.3.1) and (2.3.2) will automatically satisfy the $(m+1)$st equation; any one of the $(m+1)$ equations is redundant and can be disregarded. Similar arguments can be used to show that any m of these equations are non-redundant.

We also note that the sum of the external flows $\bar{Q}_j$ specified at the boundary nodes must satisfy (2.3.2) in order to satisfy the conservation requirement. Therefore any one of the boundary conditions of Types II and II' (flow) can be obtained from the sum of the remaining boundary conditions of these types. One of these conditions is therefore redundant.

Taking inventory of our model elements we note that the model consists of:

1. $2m$ variables representing flows Q_j and pressures P_j at the network nodes $j = 1, 2, \ldots, m$.

2. 2ℓ variables representing flows q_i through and pressure differences Δp_i across the network elements $i = 1, 2, \ldots, \ell$.

3. m equations (2.3.1) or alternatively $(m-1)$ equations (2.3.1) and one equation (2.3.2).

4. $m - k$, $(1 \leq k \leq m)$ equations $Q_j = 0$, setting node flows to zero at internal (nonboundary) nodes.

5. k equations $Q_j = \bar{Q}_j$, or $Q_j = \bar{Q}_j(P_j)$ or $P_j = \bar{P}_j$ representing boundary conditions at boundary nodes. As noted, at least one pressure must be specified and one of the specified flows is redundant.

6. ℓ equations (2.3.3) or (2.3.3') representing element characteristic relationships.

7. ℓ equations (2.3.5) representing the loop laws in the network loops.

The mathematical model thus consists of $2(m+\ell)$ equations in $2(m+\ell)$ variables. Note that parameters entering (2.3.3) or (2.3.3') are constant in each modeling problem, and may vary only from one problem to another. Therefore they are not considered as variables in this analysis.

If $m - k$ of the network nodes are "interior" because $Q_j = 0$ and because the pressure P_j is not specified there, exactly k boundary conditions must be specified at k boundary nodes. If more than k boundary conditions are specified, the problem may have no solution. If fewer than k conditions are specified, the problem is underdetermined and will have infinitely many solutions. The questions as to which k boundary conditions may be specified has not been completely answered (Gagnon and Jacoby [1975]). For example, the pressures at the network nodes (part of the solution to the modeling problem) can be determined only within an arbitrary additive constant, unless a (reference) pressure is specified at least at one boundary node. Also, the modeling problem does not make sense unless at least one external flow is specified at a boundary node. If one boundary flow is specified then the external flow at least at one other (boundary) node must be different from zero. On the other hand, it is impossible to assign arbitrary values to the flows at all boundary nodes and satisfy (2.3.2), which states that the sum of boundary node flows equals zero, at the same time. In general, certain choices of k boundary conditions may lead to unsolvable problems. Shamir and Howard [1969] give a rule for the selection of boundary conditions for water distribution networks which reduces the probability of an unsolvable problem of that type. It is possible to a certain extent to detect the unsolvability of a network problem in the course of an attempt to solve it. This possibility is further discussed in Subsection 2.3.4 below. It is important to note that solution uniqueness is an essential requirement in this modeling problem. In order to validate his design, the model user must know how the network will perform. Flows and pressures, for example, must be known to satisfy the design requirements. The modeling purpose is not achievable if more than one (possibly infinitely many) sets of flows and pressures could satisfy any stated modeling problem, i.e., any given network design and set of boundary conditions.

Birkhoff and Diaz [1956], Birkhoff [1963], and Prager [1965] consider the solvability of the problem in certain cases. We will explore the question of solution uniqueness following the outline of their discussions, considering, however, a more general problem. The problem we wish to consider first is the mixed boundary value problem, in which boundary conditions of Types I or II, or of both Types I and II, are specified at boundary nodes. The problem then is such that each network node is in one of the three subsets:

1. The subset S_p of boundary nodes at which only pressures are specified $P_j = \bar{P}_j$.

2. The subset S_f of interior and boundary nodes at which only flows are specified: either $Q_j = 0$ (interior node) or $Q_j = \bar{Q}_j$ (boundary node). Notice that all interior nodes are, by definition, in this subset.

3. The subset S_{pf} of boundary nodes at which both pressures and flows are specified, i.e., $P_j = \bar{P}_j$ and $Q_j = \bar{Q}_j$.

Notice that the mixed boundary value problem under consideration includes as special cases:

1. Boundary value problem of Type I, in which only pressure boundary conditions are specified. This occurs if S_{pf} is an empty set and S_f contains only interior nodes at which $Q_j = 0$.

2. Boundary value problem of Type II, in which only flow boundary conditions are specified. This occurs if S_p and S_{p_f} are empty sets. A reference pressure has to be specified to completely determine the solution to this problem.

3. Mixed boundary value problem of Type I and II, in which pressure and flow boundary conditions are specified, but not both at any one node. This occurs if S_{p_f} is an empty set.

To prove that the solution to the stated problem is unique, assume the existence of two different solutions to the same problem. Both solutions must satisfy the same boundary conditions. Denote these solutions by:

$$P_j^{(1)}, Q_j^{(1)}, j = 1, 2, \ldots, m \quad \text{and} \quad \Delta p_i^{(1)}, q_i^{(1)}, i = 1, 2, \ldots, \ell$$
$$P_j^{(2)}, Q_j^{(2)}, j = 1, 2, \ldots, m \quad \text{and} \quad \Delta p_i^{(2)}, q_i^{(2)}, i = 1, 2, \ldots, \ell$$

The sum

$$\sum_{j=1}^{m} (P_j^{(1)} - P_j^{(2)})(Q_j^{(1)} - Q_j^{(2)}) = 0 \qquad (2.3.6)$$

vanishes, because the first factor vanishes for all nodes in subset S_p (at which P_j are prescribed as boundary conditions), the second factor vanishes for all nodes in subset S_f (at which Q_j are prescribed) and both factors vanish for nodes in subset S_{p_f}. By postulate each network node is in one of these subsets. The fact that the sum (2.3.6) vanishes is the basis for the uniqueness proof, the outline of which follows.

It can be shown that every solution P_j, Q_j, Δp_i, and q_i must satisfy

$$\sum_{j=1}^{m} P_j Q_j = \sum_{i=1}^{\ell} \Delta p_i q_i \qquad (2.3.7)$$

This is true because of (2.3.1) and (2.3.5) which we restate here for convenience:

$$Q_j = -\sum_{i=1}^{\ell} \delta_{ij} q_i \qquad (2.3.1)$$

$$\Delta p_i = -\sum_{j=1}^{m} \delta_{ij} P_j \qquad (2.3.5)$$

To prove (2.3.7) we use these equations as follows:

$$\sum_{j=1}^{m} P_j Q_j = -\sum_{j=1}^{m} P_j \sum_{i=1}^{\ell} \delta_{ij} q_i = -\sum_{j=1}^{m} \sum_{i=1}^{\ell} \delta_{ij} q_i P_j$$
$$= -\sum_{i=1}^{\ell} \sum_{j=1}^{m} \delta_{ij} P_j q_i = \sum_{i=1}^{\ell} \Delta p_i q_i$$

Having thus established (2.3.7) we use it in (2.3.6) and obtain

$$\sum_{i=1}^{\ell} (\Delta p_i^{(1)} - \Delta p_i^{(2)})(q_i^{(1)} - q_i^{(2)}) = 0 \qquad (2.3.8)$$

We turn our attention now to (2.3.3) and note that pressure differences are postulated to be continuous strictly increasing functions of flow. Therefore the two factors in a typical term of the sum in (2.3.8) cannot have opposite signs. It follows then that the sum can only vanish if each of its terms vanishes individually. This proves that there is at most one solution to our problem since:

$$\Delta p_i^{(1)} = \Delta p_i^{(2)}, \quad \text{and} \quad q_i^{(1)} = q_i^{(2)}, \quad \text{for } i = 1, \ldots, \ell$$

and in view of (2.3.1) and (2.3.5) then also:

$$P_j^{(1)} = P_j^{(2)}, \quad \text{and} \quad Q_j^{(1)} = Q_j^{(2)}, \quad \text{for those } j = 1, \ldots, m$$

for which P_j and Q_j are not specified as boundary conditions.

Notice that for network elements on which pressure differences are continuous increasing (but not strictly increasing) functions of the flow:

$$\Delta p_i^{(1)} - \Delta p_i^{(2)} \neq 0 \quad \text{implies} \quad q_i^{(1)} = q_i^{(2)}$$

and

$$q_i^{(1)} - q_i^{(2)} \neq 0 \quad \text{implies} \quad \Delta p_i^{(1)} = \Delta p_i^{(2)}$$

Also in this case the two factors in the summand cannot have opposite signs.

We now proceed one step further and consider boundary conditions of Types I' and II'. That is, suppose that at some nodes in subset S_p the condition $\bar{P}_j(Q_j)$ is specified and at some nodes in subset S_f the condition $\bar{Q}_j(P_j)$ is specified. It is easy to see that a necessary condition for the sum (2.3.6) to vanish, and therefore also a necessary condition for uniqueness, is that the functions $\bar{P}_j(Q_j)$ and $\bar{Q}_j(P_j)$ be continuous strictly decreasing functions of their arguments.

In summary, then, a necessary condition for solution uniqueness for our problem, is that all element characteristic relationships and all boundary conditions of Type I' or II', specified as functions be continuous and increasing/decreasing (as the case may be) functions of their arguments.

The next question that needs to be explored is under what conditions does the solution to our mathematical modeling problem exist. To illustrate this analysis we follow the argument presented by Birkhoff [1963]. He considers the mixed boundary value problem of Types I and II'. That is, all the boundary nodes are in one of the subsets:

1. The subset S_p at which only pressures are specified, i.e., $P_j = \bar{P}_j$ (boundary value problem of Type I).

2. The subset S_f at which only flows are specified as functions of pressure, i.e., $Q_j = \bar{Q}_j(P_j)$ (boundary value problem of Type II'). This includes boundary value problem of Type II, with $Q_j = \bar{Q}_j$, as a special case.

As before, we assume that the

$$q_i = G_i(\Delta p_i, \beta_i)$$

are specified, continuous increasing functions of Δp_i (β_i is a constant parameter). We also assume that $\bar{Q}_j(P_j)$ are continuous, nonincreasing functions of P_j.

Following Birkhoff [1963] we define the function

$$V(\mathbf{P}) = \sum_{i=1}^{\ell} \int_0^{\Delta p_i} G_i(s)\, ds - \sum_{j \in S_f} \int_0^{P_j} \bar{Q}_j(r)\, dr \qquad (2.3.9)$$

where s and r are variables of integration, G_i is as defined in (2.3.3'), and $\bar{Q}_j$ is the boundary condition of Type II'. For nodes at which $\bar{Q}_j = $ constant, the corresponding summand in the second term reduces to $\bar{Q}_j P_j$. The function $V(\mathbf{P})$ should be viewed as a function of the pressures at those interior and boundary nodes at which pressures are not specified as boundary conditions (i.e., at interior nodes and at boundary nodes in the set S_f). That is, if the superscript T denotes transposition

$$\mathbf{P} = [P_1, P_2, \ldots, P_j, \ldots, P_m]^T$$

where it is understood that j ranges only over interior nodes and boundary nodes in S_f. The first variation of the above expression is (denoting variations by Δ and provided that G_i and $\bar{Q}_j$ are continuous functions):

$$\Delta V = -\sum_{i=1}^{\ell} G_i(\Delta p_i, \beta_i) \sum_{j=1}^{m} \delta_{ij}\, \Delta P_j - \sum_{j \in S_f} \bar{Q}_j(P_j)\, \Delta P_j$$

where the variations ΔP_j are independent over the nodes where P_j is not given.

Birkhoff first proves that the variational equation

$$\Delta V = 0 \qquad (2.3.10)$$

is equivalent to the statement of the boundary value problem under consideration. This is so because at a boundary node in the set S_f (i.e., at which Q_j is given) the coefficient of ΔP_j vanishes if and only if the boundary condition is satisfied. At the interior nodes the coefficient of ΔP_j vanishes if and only if the remainder of the problem statement is satisfied (pressures at nodes in S_p are not involved here).

Next it is shown that $V(\mathbf{P})$ defined by (2.3.9) is a convex function of the pressures if G_i are nondecreasing and $\bar{Q}_j$ are nonincreasing. If and only if G_i are increasing functions and $\bar{Q}_j$ are decreasing then $V(\mathbf{P})$ is a strictly convex function. Under the latter hypothesis there is at most one solution to the variational problem $\Delta V = 0$ and this occurs when $V(\mathbf{P})$ has a strict minimum. If the functions G_i are nondecreasing and the functions $\bar{Q}_j$ are nonincreasing and consequently $V(\mathbf{P})$ is convex, and if these functions assume both signs, then $V(\mathbf{P})$ has a minimum. The latter condition means that some of the G_i and $\bar{Q}_j$ are positive while others are negative, and it is necessary to avoid a saddle point. It follows that a (unique) solution to the variational problem $\Delta V = 0$ exists under the conditions assumed satisfied by G_i and $\bar{Q}_j$. Therefore, and because of the problem equivalence, also a (unique) solution to the boundary value problem under consideration exists. This concludes our illustration of an analysis of the solution existence of the network flow problem. The problem is now ready for solution.

2.3.4 Solution Process Development: The Implicit Loop Procedure

In this subsection we develop the first stage of the solution process for the modeling problem of steady nonlinear network flow.

The nonlinearity of the element characteristic functions implies that the solution to the modeling problem generally cannot be found by a direct, finite, computational process. Thus, iterative procedures are used whereby estimates of the variables are successively refined until all of the mathematical model equations are satisfied to within some specified tolerance. The degree to which the estimated variables satisfy any particular equation is measured by its residual—the difference between the two sides of the equation.

To maximize the efficiency and accuracy of the iterative computational process the original mathematical modeling problem is usually reformulated. The objective of the reformulation is to use some of the model equations for elimination of variables and thereby to reduce the size of the nonlinear problem that must be solved by iteration. Computational efficiency of the solution method is better for smaller problems, except when sparse matrix techniques are used. Two types of computational methods for problem size reduction are available. The first is based on graph theory. A discussion can be found in the book by Blackwell [1968]. Articles by Epp and Fowler [1970] and Kesavan and Chandrashekar [1972] show how to apply this approach to the particular network flow problem under consideration here. Another relatively simple approach due to Gagnon and Jacoby [1975], that has been successful in several applications, is described here. (Substantial portions of Subsections 2.3.4 and 2.3.5 have been reproduced by permission from this article.)

As with other network analysis methods, we are interested here in a systematic procedure for simplifying the mathematical model by eliminating variables. Thus, for example, the system of equations

$$x_1 = f(x_2)$$
$$x_2 = g(x_1)$$

(2.3.11)

where f and g are nonlinear functions of their arguments, could be solved (assuming a solution exists) by an iterative procedure applied to

$$R_1 = x_1^* - f(x_2^*)$$
$$R_2 = x_2 - g(x_1^*)$$

where asterisks denote estimates. That is, the system could be solved by successively refining x_1^* and x_2^* until R_1 and R_2 are negligible. Following the usual notion of elimination, the solution can be obtained from a single estimate by means of the iterative procedure applied to

$$R = x_2^* - g[f(x_2^*)]$$

Having a suitable value for x_2, x_1 is evaluated by means of the first equation in the given system (2.3.11). Alternatively, let $\hat{x}_1$ and $\hat{x}_2$ denote computed values which are dependent upon the estimate x_2^*, then the sequence of calculations

$$\hat{x}_1 = f(x_2^*)$$
$$\hat{x}_2 = g(\hat{x}_1)$$
$$R = x_2^* - \hat{x}_2$$

yields the identical residual as the preceding elimination process. Thus, as in the previous case, the single estimate x_2^* could be successively refined using an iterative procedure until the single residual R becomes negligible. We use this rather trivial example which extends in an obvious way to an arbitrary number of equations to illustrate that the elimination process can be effected without explicitly eliminating variables.

The set of equations which yields the residual values from a like number of estimated variables is denoted here as a *computational model* to distinguish between this set and the system of equations (mathematical model) from which it was derived.

The solution process using a computational model is illustrated in Fig. 2.3.1.

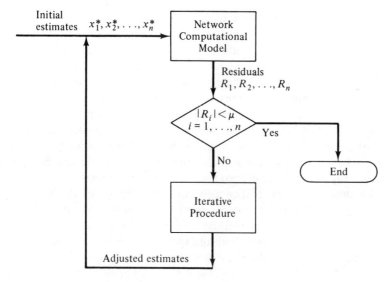

Figure 2.3.1 Network solution process.

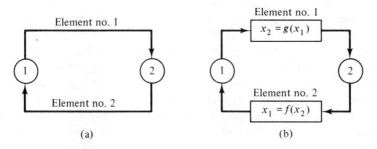

Figure 2.3.2 Simple network example: (a) graph; (b) block diagram.

The basis for the implicit loop procedure can be seen by considering Eqs. (2.3.11) which were used in the example for the elimination process. This system can be thought of as a mathematical model of a simple kind of network illustrated in Fig. 2.3.2.

Starting at node 2 in Fig. 2.3.2 the output of element number 2 can be calculated from an estimate x_2^* for x_2. The value so obtained $\hat{x}_1$ is the only input required for element number 1. Its output $\hat{x}_2$ is calculated, and the difference between x_2^* and $\hat{x}_2$ reflects the inconsistency or residual resulting from an incorrect estimate x_2^*. Thus the computational model was derived by "walking" through the network calculating outputs from inputs consisting of previously calculated quantities and, when necessary, estimated variables. The result is the same computational model which was derived previously in a more formal way.

The implicit loop method starts with a refinement of the procedure used in the foregoing example. The first goal is to establish a path through the entire network such that the number of estimates required is minimized, and a like number of residuals is produced. This path constitutes the network computational model. With some modifications which are described later, this path is established by exhausting all input/output calculations which can be performed from given, previously calculated, or previously estimated data before additional estimates are made. This process also yields a set of residuals corresponding to the initial estimates of the solution. In the computer implementation of the method, the initial estimates can be provided automatically by the computer so that the model user need not concern himself with these estimates. Notice also that the structure of the modeled network is input data and is not permanently implemented in the mathematical model. Only the implicit loop method itself is implemented in the model.

Pressure residuals arise where the (pressure) outputs of an element are incident upon a node at which the pressure has been previously calculated or specified as a boundary condition. Flow residuals arise at the output node of an element when it is the last element to be included in the path for that node. The first flow residual encountered is discarded; the equation for this node is considered the redundant flow equation (see Subsection 2.3.3 above). If it is the only flow residual, it will always be zero, otherwise it will be the negative of the sum of the remaining flow residuals. This logic guarantees that each equation in the mathematical model is either satisfied or not. If it is not satisfied, a corresponding residual is produced. For each equation that is satisfied, a variable is implicitly eliminated. It follows that the number of residuals will always equal the remaining number of variables. These remaining variables are estimated. Estimates which provide zero residuals constitute a solution to the mathematical model because all equations are satisfied when all residuals are zero. In general the initial estimates may not result in zero residuals. In this case revised estimates of the solution are produced in subsequent iterations. Residuals for these revised estimates are easily computed from the network computational model consisting of the sequence of input/output calculations in accordance with the path that has been established. This iterative process continues until it yields zero (or negligible) residuals. The estimates corresponding to these zero residuals constitute a solution satisfying the mathematical model. If the problem statement satisfies the appropriate conditions (Subsection 2.3.3) this solution is unique.

Example. Consider the simple pipe network shown in Fig. 2.3.3 where the external flows $\bar{Q}_6$ and $\bar{Q}_4$, the reference pressure $\bar{P}_1$, pipe lengths, diameters and loss coefficients are given.

Beginning at node 1 where we have $\bar{P}_1$ and $Q_1 = -\bar{Q}_4 - \bar{Q}_6$, the outputs

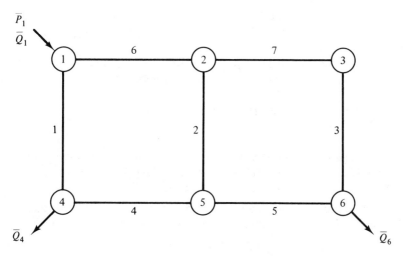

Figure 2.3.3 Pipe network diagram.

of pipes 1 and 4 in that order and of pipe 6 can be calculated from a single flow estimate q_1 or q_6. Further calculations require that one of q_2, q_5, or q_7 be estimated. Suppose q_7 is estimated, then the outputs of pipes 7, 3, 5, and 2 in that sequence can be calculated without additional estimates. At this stage every pipe has been dealt with and the path selection is completed. The path goes through the pipes in the following order: 1(1, 4), 4(4, 5), 6(1, 2), 7(2, 3), 3(3, 6), 5(6, 5), 2(5, 2); the numbers in parentheses denote the pipes' input and output nodes respectively. The estimates for the path are q_1 (or q_6), and q_7. The residuals are the difference in the output pressures of pipes 4 and 5, and the difference in the output pressures of pipes 6 and 2.

Remarks. The node flow equations are satisfied by construction at every node except node 2; i.e., every flow conservation equation is used in the process except for the equation at node 2 (the redundant equation for this case). It is easy to verify that in the above scheme the equation for node 2 of this network is always satisfied.

Since node flows are satisfied everywhere and two flow estimates which give rise to two pressure residuals are required, it follows that the procedure for this network is equivalent to a loop method. This is a method in which flows in independent loops of the network are used as the model variables.

The arbitrary decisions made in establishing the path correspond to the arbitrariness in selecting independent loops. In choosing q_1 and q_7 as the independent variables (estimated quantities) and the output pressure of pipe number 4 as the input pressure to pipe number 2, the procedure corresponds to selecting the outermost loop and the lefthand loop as the independent loops. The residual at node 5 corresponds to a pressure imbalance in the outermost loop and the residual at node 2 corresponds to a pressure imbalance in the lefthand loop.

More general networks require the additional stipulation that when an estimate is required, the estimate be made for a variable at a node which has the fewest remaining unknowns.

Stated more formally, the path selection strategy consists of the following steps:

1. Estimate all but one unknown boundary flow and compute the latter from the sum of the given and estimated boundary flows. Record the boundary flows which were estimated, then continue.

2. If no input/output calculations can be performed without estimating additional variables, go to step (4); otherwise continue.

3. Exhaust all input/output calculations which can be performed from data which have already been estimated, given, or previously calculated. Record the element numbers in a path sequence table in the order in which their outputs are calculated. Record also the input and output node for each of these elements, the residual variables and the residual values (disregarding the first flow residual), then continue.

4. If every element is in the path sequence, the path selection process is completed; otherwise continue. Find a node having a known or previously calculated pressure and the fewest number of elements which have not yet been included in the path sequence. Estimate the flow in one of these elements, record the flow variable which has been estimated, and go to step (3).

These steps produce the network computational model and the set of residuals corresponding to the initial estimates. Specifically we obtain five lists:

1. A sequence of element numbers (the path sequence) with designated input and output nodes,

2. A list of estimated boundary flows,

3. A list of estimated element flows,

4. A list of pressure residuals associated with certain element outputs,

5. A list of flow residuals associated with certain nodes.

Step (1) of the procedure uses the conservation Eq. (2.3.2) for external flows. A computer program could be implemented in a way which optionally allows all external flows to be specified. These values must, however, satisfy Eq. (2.3.2) and step (1) should, in this case, include a check to verify satisfaction of (2.3.2). Prior to the execution of the path selection steps, it should be determined if the number of specified variables (excluding one of the external flows in the case where they are all given) is equal to the number of boundary nodes. This implies that at least one node pressure is specified. As an added safeguard, it is a simple matter to determine if the number of estimates equals the number of residuals as it should. A computer implementation should also allow for the trivial case where no estimates were required and no residuals were produced. In this case, the solution is obtained by the path selection steps, and no iterative procedure needs to be applied.

The number of estimates and residuals which will be produced by this method has not been established in general as a function of the network structure and boundary conditions. Numerous experiments with both simple and complicated networks have yielded the following results:

1. The number of extimates for loop networks (networks where every element is in a loop) with a single source (Q_{in}) has always matched the number of independent loops, i.e., $m - \ell + 1$.

2. The number of estimates for loop networks with multiple sources has always matched the number of independent loops plus the number of sources in excess of one.

3. The number of estimates for networks with a tree structure with few or no loops may not be minimal. This means that one may be able to derive computational models with fewer estimates than the implicit loop method yields for these networks.

The latter situation is not likely to be serious for, e.g., water distribution networks, since it is customary to have elements in loops in order to insure continuous reliable water supply and to reduce sedimentation problems. Thus, most water distribution networks will have few, if any, nonloop elements. Furthermore, the number of estimates required by the method need not be increased by the presence of nonloop elements.

2.3.5 Solution Process Development: The Iteration Procedure

In this subsection we develop the second and final stage of the solution process for the modeling problem of steady nonlinear network flow: the iteration procedure.

One of the more successful iteration procedures for solving the type of problem we are concerned with is the Newton-Raphson method (more briefly the Newton method). Suppose the uth approximation to the independent variables (the estimates in our problem) is given by the vector

$$\mathbf{X}^{(u)} = [x_1^{(u)}, x_2^{(u)}, \ldots, x_n^{(u)}]^T$$

where the superscript T denotes transposition. Then the Newton process starts with an initial estimate $X^{(0)}$ and computes

$$X^{(u+1)} = X^{(u)} + t^{(u)}S^{(u)} \tag{2.3.12}$$

where $t^{(u)}$ is a scalar quantity (usually satisfying $0 < t^{(u)} \leq 1$) and where $S^{(u)}$ is the solution to the linear equations

$$J^{(u)}S^{(u)} = -R^{(u)}$$

$R^{(u)}$ being the vector of residuals corresponding to $X^{(u)}$, and $J^{(u)}$ being the Jacobian matrix whose typical element is given by

$$[J^{(u)}]_{ij} = \frac{\partial R_i}{\partial x_j} \qquad (i, j = 1, 2, \ldots, n)$$

evaluated at $X^{(u)}$. Symbolically the solution to the linear equations can be denoted by

$$S^{(u)} = -[J^{(u)}]^{-1} R^{(u)} \tag{2.3.13}$$

and we note that the method fails if $J^{(u)}$ is singular or sufficiently ill-conditioned. Singularity could indicate either a deterioration of the iteration procedure or an ill-posed problem (one with no solution). The vector $S^{(u)}$ is often referred to as the *direction* and $t^{(u)} S^{(u)}$ as the *step*. The iteration can be shown to converge, provided $X^{(0)}$ is close enough to the solution. We should also note that it is proper to apply the iterative procedure to the implicit loop computational model; if and when the method converges, the solution will satisfy the mathematical model equations (i.e., will be the unique solution of the modeling problem).

Quasi-Newton methods differ from Newton methods in that the matrices $J^{(u)}$ or their inverses are approximated using information obtained by the iterative process. Iteration procedures based upon quasi-Newton methods as described by Barnes [1956], Broyden [1965, 1970], and Rosen [1966] have been very successful in solving nonlinear problems of the type faced in simulation of flow networks. For example, Lam and Wolla [1972a, b] describe the application of Broyden's method [1965] to hydraulic networks. Perhaps the most attractive property of the quasi-Newton family of methods is that derivatives are not required for the type of problems we are interested in here. Starting with an approximation to the Jacobian matrix or its inverse (usually the identity matrix) these methods successively adjust the independent variables (the estimates) and refine the approximation, using the information gained in adjusting the variables. Convergence efficiency of these methods is generally very good, but like the Newton method they can fail to converge if the iteration matrix (the approximate Jacobian or its inverse) becomes singular or ill-conditioned. These effects can be offset to some extent by step size limitations as recommended by Rosen [1966], and Lam and Wolla [1972a, b] for example. This results from the fact that ill-conditioning can manifest itself by large components in the solution vector (see Wilkinson [1965], p. 255). By taking not too large a step it is possible to move away from what hopefully is a very localized phenomenon. (While it is possible to start with the identity matrix as the initial approximation to the Jacobian or its inverse, we have found in general that it is often more efficient to compute the initial approximation using finite differences.)

The process as described by Rosen [1966] is defined by

$$J^{(u+1)} = J^{(u)} + \frac{[R^{(u+1)} - R^{(u)} + J^{(u)}t^{(u)}S^{(u)}][S^{(u)}]^T}{[S^{(u)}]^T t^{(u)} S^{(u)}} \qquad (2.3.14)$$

where the superscript T denotes matrix transposition. Notice that $J^{(u+1)}$ is obtained from data which have been computed from $J^{(u)}$.

The computer implementation of our reformulated model consists of two major parts:

1. An implementation of the quasi-Newton method which computes successive estimates of the solution.

2. A residual calculation which provides the residuals corresponding to these estimates.

In addition, of course, the computer implementation will include permanent data such as:

1. Parameters α and β entering Eqs. (2.3.3) or (2.3.3'), and

2. Fluid data as necessary.

The first step of our process differs from all the others: In it the "path" is selected while residuals corresponding to the initial estimates are computed. This part of the implementation is described in Subsection 2.3.4 above. Next, control is turned over to the quasi-Newton method implementation. Thereby a sequence of new and improved estimates of the solution is computed from (2.3.12), (2.3.13), and (2.3.14). A set of current residuals, corresponding to current estimates, is required by the latter two equations. These are computed in the manner suggested in Subsection 2.3.4 and Fig. 2.3.1, in all iterations after the first. The process terminates when $t^{(u)}S^{(u)}$ in (2.3.12) is small.

This concludes the development of the solution process for the modeling problem of a steady nonlinear network flow. This model is now ready for computer implementation. In Chapter 3 (Subsection 3.2.15) we develop a general design of a computer implementation for this model.

2.3.6 Summary

In this section we have discussed a mathematical model of steady flow in pipe networks. First the prototype and the modeling purpose were defined and then a mathematical model was developed to simulate the network operation. The model is a lumped-parameter network type model; it admits several different kinds of boundary value problems, all involving a system of non-linear algebraic or transcendental equations. This mathematical model was analyzed and a summary of an argument was presented, concerning the existence of a unique solution to the different boundary value problems.

We then developed a solution process in two stages. First we formulated a procedure, called the implicit loop method, for obtaining a computational model using the model's network structure. The implicit loop method for nonlinear flow networks is very flexible and readily adaptable to a variety of networks, designs, and operating conditions. It is also an effective method for reducing the number of independent variables (estimates) and residuals to the near minimum as results with standard loop methods. This improves computational efficiency. The method is particularly well suited to digital computations, and gives a flexible model that is usable for many different network structures (modeling inputs).

The second stage of the solution process is an iteration method for solving the system of algebraic and transcendental equations in the computational model. These equations are a reformulation of the original model equations. The iterative method is basically a limited-step, modified Newton method.

2.4 A MATHEMATICAL MODEL FOR HYDROELECTRIC POWER SYSTEMS OPTIMIZATION

The subject of this section is a mathematical model of a large and complex hydroelectric power system located in the Columbia River Basin in the northwestern U.S.A. The purpose of this model is to assist in planning the long-term system operation and development. The mathematical model must relate stream flows into the system and within the system, reservoir elevations and storages, power generation and loads, and miscellaneous constraints on the operation of the system. The model must be able to compute feasible and possibly optimal operating plans for the system. An operating plan consists of reservoir storages, stream flows, and electric power generation in the system. The plan is optimal if it maximizes (or minimizes) certain measurable objectives related to the system; for example, if it maximizes the hydroelectric energy capability of the system.

An optimization model is required to achieve the modeling purpose of finding optimal operating plans. The model is a constrained optimization problem (nonlinear programming problem), of a lumped-parameter type with a tree structure. While the considered model data are related to the Columbia River Basin, the method used to formulate and solve the problem is applicable to other hydroelectric power systems. In fact the solution method which we shall present in detail is of even more general nature and interest. This example gives us an opportunity to discuss numerical approaches to large-scale optimization problems with linear and nonlinear constraints. The described solution method was implemented on a computer and used successfully to solve very large problems involving up to 6000 variables, 4000 linear equation constraints and 11,000 linear and nonlinear inequality constraints.

2.4.1   The Prototype and the Modeling Purpose

In this subsection we identify the prototype and the process considered in this modeling study; we also define the modeling purpose.

The prototype is the hydroelectric power system situated in the Columbia River Drainage Basin in the northwestern U.S.A. and in southwestern Canada. The basin extends over the states of Washington, Oregon, Idaho, and Wyoming and the province of British Columbia. When fully developed according to current plans, the system may include over 100 hydroelectric power plants. Approximately 50 of these have large water-storage reservoirs and the remaining 50 are so-called "run of the river plants" with relatively small water-storage capacity. The hydrological boundary of the system is the boundary of the Columbia River Basin. Forecasts of side flows (runoff) into the system are known. Also forecasts of outflows out of the system are known, including flows for uses such as irrigation. The net flow forecasts used for the particular modeling exercises of interest here reflect the most adverse conditions of historical record. Deterministic rather than stochastic data are used for these exercises. The power boundary of the system does not coincide with the hydrological boundary. Power loads on the system are specified as a boundary condition for modeling purposes. The system itself consists of reaches of the Columbia River and its tributaries which are interconnected in a known manner. The system includes also approximately 50 water-storage reservoirs for each of which is given the relationship between water level and volume stored. Approximately 100 power plants are considered in the system. For all these, the relationship between flow, forebay to tailwater head difference, and power generation is specified.

The purpose of the mathematical model is to assist in long-term operational and developmental planning for the prototype system. Specific model uses include:

1. Determination of seasonal water storage and power generation plans.

2. Scheduling of power plant maintenance.

3. Coordination between different agencies involved in power generation and consumption.

4. System development planning for timing of new installations.

A time period of approximately 5 years needs to be considered for this purpose and it is subdivided into close to 60 time intervals of $\frac{1}{2}$ or 1 month length. The modeled process is recursive: The system state in the first time interval depends on initial conditions. Therefore, initial conditions including the water volume stored in all reservoirs must be specified for each modeling exercise. The state of the system in the second time interval depends on its state at the end of the first interval, and so on. In general, the state of the system in any time interval depends on the state at the end of the previous time interval and on how it is operated in this interval. A modeling solution consists of a complete operating plan for the total planning period, i.e., for all time intervals. This plan includes a schedule of reservoir storages, stream flows, and electric power generation. The operating plan must satisfy or nearly satisfy miscellaneous constraints on the system operation: constraints on power generation, on flows, and on reservoir drafts and contents. It is also desired of the operating plan to maintain the surplus of power generation over loads as uniform as possible during the total planning period. An optimization model is required to achieve the modeling purpose of finding optimal operating plans, i.e., plans which, for example, maximize the hydroelectric energy capability of the system. A *lumped-parameter* type mathematical model is required because this type provides sufficient detail for the modeling purpose, and because data handling and computations would be prohibitive for a more refined model. The tree-like structure of the river tributary system can be used advantegeously in modeling, and so can certain other prototype features as is discussed further below.

2.4.2 Mathematical Model Development

In this subsection we develop a mathematical model for hydroelectric power system optimization. The discussion follows the outlines of articles by Jacoby and Kowalik [1971], by Gagnon et al. [1973, 1974], and by Hicks et al. [1974].

The entire streamflow period is considered to consist of $j = 1, 2, \ldots, J$

discrete time intervals. Most intervals are one month, but some intervals are half-months to provide added resolution.

With this time scaling, the daily and weekly cyclic variation in the ponds behind run-of-river plants are neglected. Storage for these plants is considered to be constant so that outflow is equal to inflow. Run-of-river plants are denoted by the index $k = 1, 2, \ldots, K$. Reservoir plants are denoted by the index $i = 1, 2, \ldots, I$.

The delay in water reaching a plant from upstream plants ranges from 0 to about 24 hours, which is negligible compared to the month and half-month intervals used. Plant inflow is therefore assumed to be the exogenous or side inflow plus the sum outflows of the immediate upstream plants. Side inflow data are given and are preadjusted for consumptive use.

Thus, plant interactions through common river systems are accounted for in the following *water conservation* equations. For run-of-river plants:

$$y_{kj} = \bar{X}_{kj} + \sum_{e \in \alpha_k} y_{ej} \qquad \begin{array}{l} k = 1, 2, \ldots, K \\ j = 1, 2, \ldots, J \end{array} \qquad (2.4.1)$$

where

y_{kj} is the average outflow from run-of-river plant k in time interval j

$\bar{X}_{kj}$ is the average side flow into plant k in time interval j

α_k is the set of indices of plants immediately upstream from plant k

For reservoir plants:

$$y_{ij} = \bar{X}_{ij} - \frac{c}{T_j}(x_{ij} - x_{i,j-1}) + \sum_{e \in \beta_i} y_{ej} \qquad \begin{array}{l} i = 1, 2, \ldots, I \\ j = 1, 2, \ldots, J \end{array} \qquad (2.4.2)$$

where

y_{ij} is the average outflow from reservoir plant i in time interval j

$\bar{X}_{ij}$ is the average side flow into plant i in time interval j

c is a unit conversion constant

T_j is the length of time interval j

x_{ij} is the water content of reservoir plant i at the end of time interval j

x_{i0} is the beginning water content of reservoir plant i

β_i is the set of indices of plants immediately upstream from plant i

The water conservation equation (2.4.1) and (2.4.2) can be written as a single matrix equation:

$$A\mathbf{y} + B\mathbf{x} + \mathbf{c} = 0 \qquad (2.4.3)$$

where the matrix A is square of order $[(I + K)J]$ and B is rectangular with $[(I + K)J]$ rows and $[IJ]$ columns. Both matrices are sparse, and the distribution of their nonzero elements is dependent upon the ordering of the elements y_{ij}, y_{kj} in the vector $\mathbf{y}$. The vector $\mathbf{c}$ contains the side flows and terms $x_{i_0}c/T_1$ ($i = 1, 2, \ldots, I$). $\mathbf{x}$ is a vector of x_{ij}'s.

Plants in the system interact electrically in the sense that they collectively supply the energy to meet demands. The *total power* produced by the system is expressed by

$$P_j = \sum_{i=1}^{I} P_{ij} + \sum_{k=1}^{K} P_{kj} \qquad j = 1, 2, \ldots, J \qquad (2.4.4)$$

where

P_{ij} is the average power produced by reservoir plant i in time interval j, and

P_{kj} is the average power produced by run-of-river plant k in time interval j.

Average transmission losses are accounted for in the system loads. Variations in these losses are negligible compared to power generation variations resulting from storage management changes.

System energy capability is defined as

$$P = \frac{1}{T} \sum_{j=1}^{J} P_j T_j \qquad (2.4.5)$$

where

$$T = \sum_{j=1}^{J} T_j$$

A storage management schedule for the system consists of the end-of-month storage content x_{ij} for each reservoir and this schedule, together with the exogenous flows and beginning reservoir contents, completely define the average operation of the system over each time interval and the system energy capability.

Up to this point we have introduced several assumptions to simplify the model. We have ignored:

1. The stochastic nature of the side flows $\bar{X}_{ij}$ and $\bar{X}_{kj}$,
2. River dynamics, and
3. Variations in the transmission losses.

Power generation at reservoir plants is a nonlinear function of the system variables. The nonlinear relationships have been established by semi-empirical means and are expressed in the form of tables and approximating functions.

In general reservoir plant generation can be expressed in the functional form

$$P_{ij} = P_{ij}(y_{ij}, x_{ij}, x_{i, j-1}, x_{d_i j}, x_{d_i, j-1}) \quad \begin{aligned} i &= 1, 2, \ldots, I \\ j &= 1, 2, \ldots, J \end{aligned} \quad (2.4.6)$$

where d_i is the reservoir immediately downstream of plant i. Contents of d_i appear in this equation only if the tailwater elevation in plant i depends on downstream reservoir elevation. Power generated by a run-of-river plant is a function of flow only,

$$P_{kj} = P_{kj}(y_{kj}) \quad \begin{aligned} k &= 1, 2, \ldots, K \\ j &= 1, 2, \ldots, J \end{aligned} \quad (2.4.7)$$

and is obtained from a table, one for each plant.

We will now define the *optimization objective* and *constraints*. The optimization objective is to find a storage management schedule which maximizes the system energy capability as defined by (2.4.5), with acceptable uniformity in the surplus of power over load for each time interval. This objective can be expressed as

$$\min_{\mathbf{x}, \mathbf{y}} \left\{ F(\mathbf{x}, \mathbf{y}) = D + w \sum_{j=1}^{J} (D_j - D)^2 \right\} \quad (2.4.8)$$

where

$$D_j = L_j - P_j$$

$L_j = $ Total system load for interval j

$$D = \frac{1}{T} \sum_{j=1}^{J} D_j T_j$$

and where $w > 0$ is a suitable weight. In order to be realizable and also to satisfy multipurpose stream use requirements, the storage management schedule should satisfy the water conservation equations (2.4.3) and the following inequality constraints for $i = 1, 2, \ldots, I, j = 1, 2, \ldots, J$ and $k = 1, 2, \ldots, K$:

1. Upper and lower bounds on reservoir plant outflows

$$y_{ij}^{\min} \leq y_{ij} \leq y_{ij}^{\max} \quad (2.4.9)$$

where the upper bound is a nonlinear tabulated function of the reservoir contents for a certain subset of plants,

2. Lower bounds for run-of-river plant outflows

$$y_{kj}^{\min} \leq y_{kj} \tag{2.4.10}$$

3. Upper and lower bounds on reservoir contents

$$x_i^{\min} \leq x_{ij} \leq \hat{x}_{ij} \leq x_i^{\max} \tag{2.4.11}$$

where $x_i^{\min}$ and $x_i^{\max}$ are physical limits on the reservoir contents and where $\hat{x}_{ij}$ is a desired flood control level which can be violated to some extent,

4. Upper bounds on the decrease in reservoir water elevations

$$E_{i,j-1} - E_{ij} \leq T_j \, \Delta E_i^{\max} \tag{2.4.12}$$

where E_{ij} is a given nonlinear function of the contents x_{ij} and where $\Delta E_i^{\max}$ is a given constant,

5. Upper bounds on plant generation

$$\begin{aligned} P_{ij} &\leq 0.85 P_{ij}^{\text{peak}} \\ P_{kj} &\leq 0.85 P_{kj}^{\text{peak}} \end{aligned} \tag{2.4.13}$$

where P_{ij}^{peak} is a nonlinear function of the arguments for P_{ij} as in expression (2.4.6).

The bounds on reservoir draft expressed by (2.4.12) are imposed to prevent excessive soil erosion around the reservoirs. The bounds (2.4.13) are imposed because of decreases in generation efficiency above the bounds. These latter bounds also provide generation flexibility to meet the daily and weekly energy demands which fluctuate about the average loads.

The optimization problem can now be summarized as follows:

$$\underset{\mathbf{x},\mathbf{y}}{\text{minimize}} \; F(\mathbf{x}, \mathbf{y}) \tag{2.4.14}$$

subject to

$$A\mathbf{y} + B\mathbf{x} + \mathbf{c} = 0 \tag{2.4.15}$$

and

$$\mathbf{y}^{\min} \leq \mathbf{y} \leq \mathbf{y}^{\max}(\mathbf{x}) \tag{2.4.16}$$

$$\mathbf{x}^{\min} \leq \mathbf{x} \leq \hat{\mathbf{x}} \leq \mathbf{x}^{\max} \tag{2.4.17}$$

$$\mathbf{h}(\mathbf{x}) \leq \Delta \mathbf{E}^{\max} \tag{2.4.18}$$

$$P(\mathbf{x}, \mathbf{y}) \leq 0.85 \, \mathbf{P}^{\text{peak}}(\mathbf{x}, \mathbf{y}) \tag{2.4.19}$$

Problem (2.4.14)–(2.4.19) is a nonlinear programming problem in which the objective function is nonlinear, and the constraints are a mixture of linear and nonlinear constraints.

2.4.3 Mathematical Modeling Problem Analysis: I

In this subsection we conduct the first stage of analysis and reformulation of the mathematical modeling problem developed for hydroelectric power system optimization.

Our first observation is that constraints (2.4.15)–(2.4.19) do not necessarily define a nonempty feasible region. In fact, results of simulation studies that preceded the project described here, indicate that a feasible point (x, y) may not even exist. Thus, strictly speaking, the problem as stated may not have a solution. However, we can identify two types of constraints in this formulation; we will call them hard and soft constraints. Constraints which are derived from physical properties of the system and cannot be violated under any circumstances are hard. The remaining constraints are soft, because they express desirable operating conditions which can be violated to some extent. The hard constraints include (2.4.15) and (2.4.17). All of the remaining constraints are soft. We can transfer a portion of the constraints (soft constraints) to the objective function in the form of penalty functions. The purpose of this is to impose an increasing penalty on the objective function as the constraint violation increases. All our soft constraints can be represented by inequalities of the form $g(z) \leq 0$. We would like to use a function $\psi(g)$, defined and continuous on $-\infty < g < \infty$ and such that $\psi(g) \to \infty$, when $g \to \infty$ and $\psi(g) = 0$ when $g < 0$. Several functions may be used for this purpose and our choice is

$$\psi(g(x)) = [\max(0, g(x))]^2 \qquad (2.4.20)$$

since it is convenient to use differentiable penalty functions.

For additional information on these transformation methods the reader may consult Ryan [1974]. The augmented objective function is now:

$$f(x, y) = F(x, y) + \sum_{\ell} w_{\ell}\psi(g_{\ell}) \qquad (2.4.21)$$

where ℓ ranges over all soft constraints and where $w_{\ell} > 0$ are suitable weights.

The new problem to consider is:

$$\min_{x, y} f(x, y) \qquad (2.4.22)$$

subject to

$$Ay + Bx + c = 0 \qquad (2.4.23)$$

$$x^{\min} \leq x \leq x^{\max} \qquad (2.4.24)$$

This problem is still large in size but it involves only linear constraints (2.4.23) and simple bounds on x (2.4.24). Problems of this kind usually present a nontrivial computational task due to substantial memory requirements and large computing time needed to find an accurate solution. Before we pursue our analysis further we will briefly review optimization methods applicable to linearly constrained optimization problems.

2.4.4 Linearly Constrained Minimization Problems

This subsection provides background material on linearly constrained minimization problems. This background is required for the second stage of analysis of our mathematical modeling problem.

Linearly constrained problems have received considerable attention among researchers. There are several reasons explaining this interest. Firstly, many real-life applications, particularly in the area of operations research involve linearly constrained problems. Secondly, several unconstrained optimization methods can be extended to handle linear constraints. Thirdly, it is a natural step toward solving more difficult problems with nonlinear constraints. To explain the main approaches to linearly constrained problems let us consider the case with linear equations

$$\min_{\mathbf{x}} f(\mathbf{x}) \tag{2.4.25}$$

subject to

$$C^T\mathbf{x} = \mathbf{b} \tag{2.4.26}$$

where C is an $n \times m$ matrix with $m < n$ and $\mathbf{x}$ is an $n \times 1$ vector. Suppose that the point $\mathbf{x}^i$ is feasible, then the subsequent approximation to the minimum, $\mathbf{x}^{i+1} = \mathbf{x}^i + \mathbf{d}^i$ is feasible if

$$C^T\mathbf{d}^i = 0 \tag{2.4.27}$$

Relation (2.4.27) means that $\mathbf{d}^i$ is orthogonal to the constraint normals (rows of C^T). Assume that close to $\mathbf{x}^i$ the function $f(\mathbf{x})$ can be approximated by a second-order Taylor expansion with a positive definite Hessian matrix $G(\mathbf{x}^i)$

$$f^{i+1} = f^i + (\nabla f^i)^T\mathbf{d}^i + \tfrac{1}{2}(\mathbf{d}^i)^T G^i \mathbf{d}^i \tag{2.4.28}$$

where f^i means $f(\mathbf{x}^i)$ and ∇f^i is the gradient vector of $f(\mathbf{x})$ at $\mathbf{x}^i$. We can now consider an approximated problem,

$$\min_{\mathbf{d}^i} \{(\nabla f^i)^T\mathbf{d}^i + \tfrac{1}{2}(\mathbf{d}^i)^T G^i \mathbf{d}^i\} \tag{2.4.29}$$

subject to

$$C^T\mathbf{d}^i = 0 \tag{2.4.30}$$

whose solution will give us a feasible search direction at $\mathbf{x}^i$, and such that

$$f(\mathbf{x}^i + \lambda^i \mathbf{d}^i) < f(\mathbf{x}^i) \qquad (2.4.31)$$

for some λ^i.

Suppose that we know a set of linearly independent vectors spanning the space orthogonal to the rows of C^T. Let it be the matrix Z, which is of order $n(n - m)$. The columns of Z are the spanning vectors. The vector $\mathbf{d}^i$ can be expressed in terms of the columns of Z,

$$\mathbf{d}^i = Z\mathbf{y} \qquad (2.4.32)$$

where $\mathbf{y}$ is unique. Introducing (2.4.32) to (2.4.29) we get

$$\min_{\mathbf{y}} \{(\nabla f^i)Z\mathbf{y} + \tfrac{1}{2}\mathbf{y}^T(Z^T G^i Z)\mathbf{y}\} \qquad (2.4.33)$$

which is an unconstrained problem.

The quadratic function (2.4.33) has the unique minimum

$$\mathbf{y} = -(Z^T G^i Z)^{-1} Z^T \nabla f^i \qquad (2.4.34)$$

The search direction $\mathbf{d}^i$ is obtained by substituting (2.4.34) in (2.4.32)

$$\mathbf{d}^i = -Z(Z^T G^i Z)^{-1} Z^T \nabla f^i \qquad (2.4.35)$$

and can be used to find

$$\mathbf{x}^i + \lambda^i \mathbf{d}^i = \mathbf{x}^{i+1}$$

such that $f(\mathbf{x}^{i+1}) < f(\mathbf{x}^i)$.

This concludes the i-iteration of the optimization procedure. The method described is a direct extension of Newton's method for unconstrained problems. In the absence of linear constraints we set $Z = I$ (unit matrix) and get the unconstrained Newton search direction

$$\mathbf{d}^i = -(G^i)^{-1} \nabla f^i \qquad (2.4.36)$$

In the numerical implementation the matrix $Z^T G^i Z$ may have to be modified to secure its positive definiteness.

It is possible to extend this method so that it can handle a mixed set of linear equations and inequalities. For the detailed explanations of this method the reader is referred to Gill and Murray [1974]. These extensions of Newton's method have several advantages:

1. The methods are effective for problems with a highly nonlinear objective function,

2. They are efficient in terms of the total number of iterations required to find a solution, and

3. They are robust.

On the other hand these methods require evaluation and storage of the Hessian matrix G^i. It is possible to avoid computation of the second-order derivatives by using the extended quasi-Newton methods which use approximations to G^i. The quasi-Newton methods require only the computation of the first derivative vectors ∇f^i. However, the need to store the approximate matrices is retained. For large-scale problems we may prefer to avoid storage of Hessians or their approximations and the methods that can accomplish this become more attractive. An example of such a method is a modified Fletcher-Reeves [1964] conjugate-gradient method which in its original form was designed for unconstrained problems. Let us recall that

$$\mathbf{v}_p = P\mathbf{v} = (I - C(C^T C)^{-1} C^T)\mathbf{v} \qquad (2.4.37)$$

is a projection of the vector $\mathbf{v}$ on to the space spanned by the vectors orthogonal to the rows of C^T. Using the matrix P we can project the search directions generated by the Fletcher-Reeves method and search feasible approximations to the minimum solution of $f(\mathbf{x})$.

The reduced gradient method offers a different way to generate a feasible sequence of approximations to the solution of problem (2.4.25)–(2.4.26). Assuming that the matrix C is full rank we can partition C^T and $\mathbf{x}$ as follows:

$$C^T = (A, B) \quad \text{and} \quad \mathbf{x}^T = (\mathbf{x}_1^T, \mathbf{x}_2^T)$$

where A is a nonsingular, $n \times n$ matrix.

The equation $C^T\mathbf{x} = \mathbf{b}$ becomes

$$A\mathbf{x}_1 + B\mathbf{x}_2 = \mathbf{b} \qquad (2.4.38)$$

which can be solved for $\mathbf{x}_1$:

$$\mathbf{x}_1 = A^{-1}(\mathbf{b} - B\mathbf{x}_2) \qquad (2.4.39)$$

We can interpret (2.4.39) as an attempt to eliminate m variables from the original problem. The optimization process is now reduced to the space of the variables $\mathbf{x}_2$, and $\mathbf{x}_1$ is computed from (2.4.39). Our problem (2.4.22)–(2.4.24) lends itself to the application of this method. We now return to the second part of the mathematical model analysis.

2.4.5 Mathematical Modeling Problem
Analysis: II

This subsection contains the second stage of analysis of the mathematical modeling problems involved in hydroelectric power system optimization.

Problem (2.4.22)–(2.4.24) is an excellent illustration of a favorable constraint structure. Firstly, the matrices A and B are very sparse since there are only few nonzeros in each row. Secondly, simple forms of A and B result when the assigned plant numbers increase as one proceeds downstream (or upstream). Here we take advantage of the tree structure of the hydroelectric network. Such a numbering system forces the matrices A and B to become lower triangular and lower trapezoidal, respectively. To show that this indeed happens consider the first upstream plant on the river, $i = 1$. This gives

$$y_{11} = \bar{X}_{11} - \frac{C}{T_{11}}(x_{11} - x_{10})$$

in which the constants C and T, the initial condition x_{10} and the side flow $\bar{X}_{11}$ are all given. Thus we obtain a linear equation giving y_{11} in terms of x_{11}. Now suppose that plant $i = 2$ is the first upstream plant on another tributary river. Similar analysis shows that y_{21} can be obtained in terms of x_{21} and given data. Next take plant $i = 3$, assumed to be on the river junction of the two tributaries, just downstream from plants $i = 1$ and $i = 2$. Consider (2.4.2) as applied to this plant:

$$y_{31} = \bar{X}_{31} - \frac{C}{T_1}(x_{31} - x_{30}) + y_{11} + y_{21}$$

Again C, T, x_{30}, and $\bar{X}_{31}$ are known. Furthermore y_{11} and y_{21} are known in terms of x_{11} and x_{21}, respectively. It follows that y_{31} can be computed. Continuing through the remaining plants in the same fashion and then repeating for the second time period with $j = 2$ and so on we see how the matrices A and B become lower triangular and lower trapezoidal, respectively. System (2.4.23) can be easily solved for $\mathbf{y}$ by the forward substitution process. The rows of A and B can be generated as needed from the problem data and essentially no additional storage is needed to solve equations (2.4.23).

The matrix A is unit triangular (therefore nonsingular) and the solution $\mathbf{y}$ can be symbolically written as:

$$\mathbf{y} = -A^{-1}(B\mathbf{x} + \mathbf{c}) \tag{2.4.40}$$

and the reformulated optimization problem is

$$\min_{\mathbf{x}} f(\mathbf{x}, \mathbf{y}(\mathbf{x})) \qquad (2.4.41)$$

subject to

$$\mathbf{x}^{\min} \leq \mathbf{x} \leq \mathbf{x}^{\max} \qquad (2.4.42)$$

This variable reduction can in principle be applied to any problem with a set of linear equation constraints such as (2.4.23). However, if A is very large and randomly sparse then a good solution method has to use special sparse matrix techniques to avoid excessive growth of nonzeros during the Gaussian elimination process. In our case the Gaussian elimination process is not necessary since A is triangular. Another advantage of having A in triangular form is the good accuracy of the solution $\mathbf{y}$.

Function (2.4.41) can be minimized by any unconstrained optimization algorithm provided that simple bounds (2.4.42) are enforced. The candidate methods include:

1. Newton-type methods (Murray [1972]),
2. Quasi-Newton methods (Broyden [1972]),
3. Conjugate gradient methods (Fletcher [1972]).

Methods (1) and (2) are superior in terms of convergence properties and over-all performance, but require storage and manipulation of large matrices of order n for problems with n independent variables. On the other hand, methods (3) require more iterations to converge but have the advantage of simplicity and very modest storage requirement. We have selected the conjugate gradient method of Fletcher and Reeves [1964] which has a history of successful practical applications. The method requires calculation of the gradient of f with respect to $\mathbf{x}$ which is given by

$$\nabla f_x = \frac{\partial f}{\partial \mathbf{x}} + \left(\frac{\partial \mathbf{y}}{\partial \mathbf{x}}\right)^T \frac{\partial f}{\partial \mathbf{y}} \qquad (2.4.43)$$

where from (2.4.40)

$$\left(\frac{\partial \mathbf{y}}{\partial \mathbf{x}}\right)^T = -B^T(A^T)^{-1} \qquad (2.4.44)$$

The vector ∇f_x can be conveniently computed in two steps:

1. Solve the linear system

$$A^T \boldsymbol{\lambda} = -\frac{\partial f}{\partial \mathbf{y}} \qquad (2.4.45)$$

for the vector $\boldsymbol{\lambda}$, and

2. Compute

$$\nabla f_x = \frac{\partial f}{\partial \mathbf{x}} + B^T \boldsymbol{\lambda} \tag{2.4.46}$$

2.4.6 Solution Algorithm

In this subsection we discuss the solution algorithm for the hydroelectric power system optimization problem.

The main steps of the solution method are as follows:

1. Start with $\mathbf{x}^0$ satisfying

$$\mathbf{x}^{\min} \leq \mathbf{x}^0 \leq \mathbf{x}^{\max}$$

and set $\ell = 0$

2. Solve

$$A\mathbf{y}^\ell + B\mathbf{x}^\ell + \mathbf{c} = 0$$

for $\mathbf{y}^\ell$.

3. Solve

$$A^T\boldsymbol{\lambda}^\ell = -\frac{\partial f}{\partial \mathbf{y}}$$

for $\boldsymbol{\lambda}^\ell$, where $\partial f/\partial \mathbf{y}$ is evaluated at $\mathbf{y}^\ell$, $\mathbf{x}^\ell$.

4. Compute

$$\nabla f_x^\ell = \frac{\partial f}{\partial \mathbf{x}} + B^T\boldsymbol{\lambda}^\ell$$

where $\partial f/\partial \mathbf{x}$ is evaluated at $\mathbf{y}^\ell$, $\mathbf{x}^\ell$.

5. Compute the Fletcher-Reeves conjugate search direction

$$\mathbf{t}^i = \begin{cases} -\nabla f_x + \dfrac{||\nabla f^\ell||^2}{||\nabla f^{\ell-1}||^2} \hat{t}^{\ell-1} & \text{for } \ell \geq 1 \\ -\Delta f_x^\ell & \text{for } i = 0 \end{cases} \tag{2.4.47}$$

where $||\mathbf{v}||^2 = \sum_r v_r^2$ for any vector $\mathbf{v}^T = (v_1, v_2, \ldots, v_n)$. We also set $\mathbf{t}^\ell = -\nabla f_x^\ell$ if $\mathbf{t}^\ell$ computed by the upper part of formula (2.4.47) is not descent (downhill) or almost orthogonal to $-\nabla f_x^\ell$.

6. Project the components of $\mathbf{t}^\ell$ onto the bounds for $\mathbf{x}$

$$\hat{t}_r^\ell = \begin{cases} 0 & \text{if } x_r^{\max} - x_r^\ell \leq \epsilon \text{ and } t_r^\ell > 0 \\ 0 & \text{if } x_r^\ell - x_r^{\min} \leq \epsilon \text{ and } t_r^\ell < 0 \\ t_r^\ell & \text{otherwise} \end{cases}$$

where $r = 1, 2, \ldots, IJ$ and $\epsilon > 0$.

7. Find the maximum step length $\theta^{\max}$ that can be taken from $\mathbf{x}^\ell$ in the direction $\hat{\mathbf{t}}^\ell$ without violating the $\mathbf{x}$ bounds,

$$\theta^{\max} = \min \left\{ \min_r \left[\frac{x_r^{\max} - x_r^\ell}{\hat{t}_r^\ell} \right]_{\hat{t}_r^\ell > 0}, \min \left[\frac{x_r^{\min} - x_r^\ell}{\hat{t}_r^\ell} \right]_{\hat{t}_r^\ell < 0} \right\}$$

8. Find $\theta^\ell \leq \theta^{\max}$ such that

$$f(\mathbf{x}^{\ell+1}, \mathbf{y}(\mathbf{x}^{\ell+1})) < f(\mathbf{x}^\ell, \mathbf{y}(\mathbf{x}^\ell))$$

where

$$\mathbf{x}^{\ell+1} = \mathbf{x}^\ell + \theta^\ell \hat{\mathbf{t}}^\ell$$

and where $\mathbf{y}(\mathbf{x}^\ell)$ is obtained in step (2).

9. If $\ell = IJ$ repeat from (1) with $\mathbf{x}^0 = \mathbf{x}^{\ell+1}$, increment ℓ by one and repeat the process starting with (2).

The following comments are in order:

The vectors of partial derivatives are obtained through the chain rule using slopes of the tabulated functions. For reasons unrelated to this problem, it was required that linear interpolation be used on all tables. For computational efficiency the piecewise constant slopes were therefore precomputed and stored with the tables.

The matrices A and B are introduced in the foregoing descriptions for notational convenience; explicit computer storage and manipulation with these matrices is neither necessary nor desirable. Recalling that these matrices stem from the water conservation equations (2.4.1) and (2.4.2), observe that these equations can be solved for the y_{ij} and y_{kj} quantities from given contents and side-flow data. This can be done by starting with the upstream-most plants (where β_i and α_k are empty sets) and proceeding downstream in such a way that the outflows of plants in the set β_i and α_k have been computed prior to computing the outflow of plant i and k. Thus, Eqs. (2.4.1) and (2.4.2) symbolized by the matrix equation in step (2) can be solved by an appropriate ordering of the plants, given a table of the sets β_i and α_k ($i = 1, 2, \ldots, I$; $k = 1, 2, \ldots, K$). The form of the matrix B is clear from these equations;

the simple logical operations represented by this matrix can be embedded in a computer code.

The solution to the matrix equation in step (3) can also be obtained in a simple manner. With the foregoing downstream plant ordering, the matrix A is unit lower triangular, and the process of solving for the outflows is in effect a sequence of forward substitutions. Thus the multipliers λ can be calculated through a sequence of backward substitutions. The process can be further expedited. Note that, for any row of A, the nonzero off-diagonal elements correspond to plants immediately upstream of the plant represented by the diagonal element. As a consequence, for any column of A (row of A^T), nonzero off-diagonal elements correspond to the plant immediately downstream of the diagonal element plant. Thus, the multipliers λ are easily computed from a table of immediate downstream neighbors to each plant.

Unique solutions to the foregoing equations exist by virtue of the fact that the matrix A is unit lower triangular and therefore nonsingular. Furthermore, since the off-diagonal elements of A represent the sum of the outflows of upstream plants, severe cancellation cannot occur. This implies that A is well conditioned with respect to the linear equation solutions.

Finally the question of uniqueness and convergence should be discussed. As mentioned before, problem (2.4.14)–(2.4.19) may not have a feasible solution, i.e., a pair of $(\mathbf{x}, \mathbf{y})$ satisfying all constraints (2.4.15)–(2.4.19). The reformulated problem (2.4.22)–(2.4.24) does have feasible solutions and we can search for a minimum of $f(\mathbf{x}, \mathbf{y})$ which depends on weights related to the soft constraints. Due to the complicated nature of the nonlinear functions involved in the formulation of $f(\mathbf{x}, \mathbf{y}(\mathbf{x}))$, it is not possible to determine whether or not the function is convex. Consequently, we do not know whether a global minimum exists for any fixed set of weights. The most that can be expected of any optimization algorithm is that it will be able to improve upon the objective function value corresponding to a given initial estimate of the variables.

Not much can be said about the method's convergence to a solution of this problem. A minimum of the problem should satisfy the necessary condition that the projected vector ∇f_x (onto the simple bounds) is zero or sufficiently close to zero. In the problem at hand, however, the given data are such that the first derivatives of the power functions are only piecewise continuous and this stopping criterion is not appropriate. We have used an alternative convergence criterion

$$f(\mathbf{x}^\ell, \mathbf{y}(\mathbf{x}^\ell)) - f(\mathbf{x}^{\ell+1}, \mathbf{y}(\mathbf{x}^{\ell+1})) \leq \epsilon_1$$

and

$$\|\mathbf{x}^{\ell+1} - \mathbf{x}^\ell\| \leq \epsilon_2$$

where ϵ_1 and ϵ_2 are suitably small numbers. Thereby the iteration is terminated when changes in the objective function and in the independent variables **x** become sufficiently small and further progress cannot be made.

This concludes our discussion of the solution for the hydroelectric power system optimization model. The model is now ready for computer implementation.

2.4.7 Summary

A mathematical model of a hydroelectric power system is discussed in this section. First the prototype and the modeling purpose are defined and then a mathematical model is developed. The prototype is the Columbia River Basin hydroelectric power system and the purpose of the model is operational and developmental planning. The model is a lumped-parameter model with a tree structure. The mathematical problem involved in it is a constrained nonlinear optimization problem. The model is analyzed and reformulated, using two main problem features. The relative "softness" of some of the constraints is used to formulate a problem consisting of a penalty function which is minimized subject to a set of linear constraints and simple bounds imposed on problem's variables. Then the tree-like structure of the model is used to reduce the number of independent variables. The method's implementation was also discussed.

The problem presented is a good example of a large-scale optimization model that has been successfully solved and implemented in practice.

2.5 REVIEW OF EXAMPLES

The purpose of this section is to retrace the development of our three case studies. This material was covered in Sections 2.2–2.4. We addressed three prototype phenomena: steady flow in a two-dimensional medium, steady nonlinear network flow, and hydroelectric power system optimization. We pursued the development of a mathematical model for each of these prototype phenomena, from the beginning of the process through the analytical stage and to the point at which the models are ready for computer implementation. This development includes: prototype identification and statement of the modeling problem, mathematical model definition and analysis, mathematical problem analysis and reformulation and solution algorithm development.

Table 2.5.1 retraces this development of the three case studies.

Table 2.5.1 Summary of Case Studies in Sections 2.2–2.4

	FLOW IN TWO-DIMENSIONAL MEDIUM (Section 2.2)	NONLINEAR NETWORK FLOW (Section 2.3)	HYDROELECTRIC POWER SYSTEM OPTIMIZATION (Section 2.4)
Prototype	Two-dimensional flow-conducting continuous medium	Network of interconnected pipes and fittings	Systems of interconnected rivers and hydroelectric power plants
Modeled phenomenon	Steady flow across boundary and within medium	Steady flow through network	Water flow and storage and power generation in system as a function of time
Modeling purpose	Obtain description of flow field	Find flows and pressures everywhere in network	Plan water flow and storage in the system to maximize electric power generation
Model type	Distributed-parameter in two spatial dimensions, steady-state, deterministic parameters	Interconnected lumped-parameter elements, steady-state deterministic parameters	Interconnected lumped-parameter elements, quasi-steady-state, deterministic parameters
Component model definition	Conservation of flow in dissipator element (distributed)	Conservation of flow network element (lumped)	Conservation of flow and power generation in system element (lumped)
Model boundary definition	Three types of boundary models corresponding to three types of boundary behavior, deterministic data (potential function, potential gradient or mix)	Three types of boundary models corresponding to three types of boundary behavior, deterministic data (pressure, flows or mix)	Inflow and initial water storage conditions specified, deterministic data
Model structure	Two-dimensional, distributed-parameter model	Complex, network model of inter-connected lumped elements	Complex, tree model of inter-connected lumped elements

Table 2.5.1 *(cont.)*

	FLOW IN TWO-DIMENSIONAL MEDIUM (Section 2.2)	NONLINEAR NETWORK FLOW (Section 2.3)	HYDROELECTRIC POWER SYSTEM OPTIMIZATION (Section 2.4)
Mathematical modeling problem definition	Boundary value problem involving partial differential equation	Boundary value problem involving nonlinear algebraic and transcendental equations	Nonlinearly constrained nonlinear optimization problem
Mathematical problem solvability	Solvable in general under certain conditions	Solvable in general under certain conditions	Only practical solution obtainable
Mathematical problem reformulation	Set of linear finite-difference equations	No reformulation	Linearly constrained nonlinear optimization problem
Mathematical problem solution method	Numerical solution of set of linear algebraic equations	Development of equation set corresponding to network structure and numerical solution of set of nonlinear algebraic and transcendental equations	Numerical solution of linearly constrained nonlinear optimization problem
Remarks	Alternative model developed—lumped, finite-element model—leading to same mathematical problem	Please refer to Subsection 3.2.15 for a general design of a computer implementation of this model	

REFERENCES

BARNES, J. G. P., "An Algorithm for Solving Nonlinear Equations Based on the Secant Method," *Comp. J.*, vol. 8, no. 1, 66, 1965.

BIRKHOFF, G., and J. B. DIAZ, "Nonlinear Network Problems," *Quarterly of Applied Math.*, vol. XIII, no. 4, 431, 1956.

BIRKHOFF, G., "A Variational Principle for Nonlinear Networks," *Quarterly of Applied Math.*, vol. XXI, no. 2, 160, 1963.

BLACKWELL, V. A., *Mathematical Modeling of Physical Networks*. New York: Macmillan, 1968.

BROWN, K., K. HILLSTROM, M. MINKOFF, L. NAZARETH, J. POOL, and B. SMITH, "A MINPACK Project Progress Report," *Argonne National Laboratory, Report TM-255*, 1975.

BROYDEN, C. G., "A Class of Methods for Solving Nonlinear Simultaneous Equations," *Math. of Comp.*, vol. 19, no. 92, 577, 1965.

BROYDEN, C. G., "The Convergence of Single-Rank Quasi-Newton Methods," *Math. of Comp.*, vol. 24, no. 110, 365, 1970.

BUNCH, J. R., and B. N. PARLETT, "Direct Methods for Solving Symmetric Indefinite Systems of Linear Equations," *SIAM J. Numer. Anal.*, vol. 8, no. 4, 1971.

CATE, E. G., A. M. ERISMAN, P. LU, and R. M. SOUTHALL, "A User Oriented Multi-Level Math Library," *Mathematical Software II, ACM-SIAM Conference*, Purdue University, 1974.

CHARTRES, B. and R. STEPLEMAN, "A General Theory of Convergence for Numerical Methods," *SIAM J. Numer. Anal.*, vol. 9, no. 3, 476, 1972.

CODY, W. J., "The Construction of Numerical Subroutine Libraries," *SIAM Review*, vol. 16, 36, 1974.

COLLATZ, L., *The Numerical Treatment of Differential Equations*. Berlin: Springer Verlag, 1960.

DAHLQUIST, G., and A. BJÖRK, *Numerical Methods*. Englewood Cliffs, N.J.: Prentice-Hall, 1974.

EPP, R., and A. G. FOWLER, "Efficient Code for Steady State Flows in Networks," *Journal of the Hydraulics Division*, ASCE, vol. 96, no. HY1, 43, Proc. Paper 7002, 1970.

FLETCHER, R., and C. M. REEVES, "Function Minimization by Conjugate Gradients," *Computer Journal*, vol. 7, 149, 1964.

FORD, B., and D. K. SAYERS, "Developing a Single Numerical Algorithms Library for Different Machine Ranges," *ACM Trans. on Mathematical Software*, vol. 2, 115, 1976.

FOX, L., and D. F. MAYERS, *Computing Methods for Scientists and Engineers*. Oxford: Clarendon Press, 1968.

GAGNON, C. R., R. H. HICKS, S. L. S. JACOBY and J. S. KOWALIK, "A Mathematical Model and a Digital Computer Program for Automatic Hydroelectric Power System Critical Period Analysis," Report BCS-G0216, Boeing Computer Services, Inc., Seattle, Wash., 1973.

GAGNON, C. R., R. H. HICKS, S. L. S. JACOBY and J. S. KOWALIK, "A Nonlinear Programming Approach to a Very Large Hydroelectric System Optimization," *Mathematical Programming*, vol. 6, no. 1, 28, 1974.

GAGNON, C. R. and S. L. S. JACOBY, "Computer Simulation of Water Distribution Networks," *Transportation Engineering Journal*, ASCE, vol. 101, no. TE3, 553, 1975.

GARABEDIAN, P. R., *Partial Differential Equations*. New York: Wiley, 1964.

GEORGE, J. A., "Block Eliminations on Finite Element Systems of Equations," in *Sparse Matrices and Their Applications*, D. J. Rose and R. A. Willoughby (eds.). New York: Plenum Press, 1972.

GIEDT, W. H., *Principles of Engineering Heat Transfer*. Princeton, N.J.: D. Van Nostrand Co., 1957.

GILL, P. E., and W. MURRAY (eds.), *Numerical Methods for Constrained Optimization*. New York: Academic Press, 1972.

HICKS, R. H., C. R. GAGNON, S. L. S. JACOBY and J. S. KOWALIK, "Large Scale, Nonlinear Optimization of Energy Capability for the Pacific Northwest Hydroelectric System," *Transactions on Power Apparatus and Systems*, IEEE, vol. PAS-93, no. 5, 1604 (paper T 74 110–3), 1974.

HIMMELBLAU, D. M. (ed.), *Decomposition of Large Scale Problems*. New York: American Elsevier Publishing Co., 1973.

HOPPER, M. J., "Harwell Subroutine Library," Atomic Energy Res. Establishment, *Report AERE-R 7477*, 1973.

HOUSEHOLDER, A. S., *The Theory of Matrices in Numerical Analysis*. Boston: Blaisdell, 1964.

JACOBY, S. L. S., and D. W. TWIGG, "Computer Solutions to Distribution Network Problems," *Proc. National Symposium on the Analysis of Water-Resource Systems*, AWRA, Denver, Colo., 167, 1968.

JACOBY, S. L. S., and J. S. KOWALIK, "Optimal Planning of Flows in Multi-Reservoir HydroPower Systems," *Proceedings of the International Symposium on Mathematical Models in Hydrology*, International Association of Hydrological Sciences, Warsaw, Poland, vol. 2, 1030, 1971.

KESAVAN, H. K., and M. CHANDRASHEKAR, "Graph Theoretic Models for Pipe Network Analysis," *Journal of the Hydraulics Division*, ASCE, vol. 98, no. HY2, 345, Proc. Paper 8709, 1972.

KING, H. W., *Handbook of Hydraulics*. New York: McGraw-Hill, 1954.

KNUTH, D. E., *The Art of Computer Programming*, Vol. 2, Semi-numerical Algorithms. Reading, Mass.: Addison-Wesley, 1969.

KOOPMAN, B. O., "Intuition in Mathematical Operations Research," *Operations Research*, vol. 25, 189, 1977.

LAM, C. F., and M. L. WOLLA, "Computer Analysis of Water Distribution Systems: Part I—Formulation of Equations," *Journal of the Hydraulics Division*, ASCE, vol. 98, no. HY3, 335, 1972a.

LAM, C. F., and M. L. WOLLA, "Computer Analysis of Water Distribution Systems: Part II—Numerical Solution," *Journal of the Hydraulics Division*, ASCE, vol. 98, no. HY3, 447, 1972b.

LASDON, L. S., and J. D. SCHOEFFLER, "Decentralized Plant Control," *ISA Trans.*, vol. 5, 1966.

MARTIN, M. S., and D. J. ROSE, *Complexity Bounds for Regular Finite Difference and Finite Element Grids*, Univ. of Denver, Rep. MS-R-7221, Denver, Colorado, 1972.

NICKEL, K., and K. RITTER, "Termination Criterion and Numerical Convergence," *SIAM J. Numer. Anal.*, vol. 9, no. 2, 277, 1972.

PEARSON, J. D., "Decomposition, Coordination, and Multilevel Systems," *IEEE Transactions on Systems Science and Cybernetics*, vol. SSC-2, no. 1, 1966.

PRAGER, W., "Problems of Network Flow," *ZAMP*, vol. 16, 185, 1965.

REID, J. K. (ed.), *Large Sparse Sets of Linear Equations*. New York: Academic Press, 1971.

RICE, J. R. (ed.), *Mathematical Software*. New York: Academic Press, 1971.

ROSE, D. J., and R. A. WILLOUGHBY (eds.), *Sparse Matrices and Their Applications*. New York: Plenum Press, 1972.

ROSEN, E. M., "A Review of Quasi-Newton Methods in Nonlinear Equation Solving and Unconstrained Optimization," *Proceedings of the 21st National Conference*, ACM, 37, 1966.

ROSENBLUETH, A., and N. WIENER, "The Role of Models in Science," *Philosophy of Science*, vol. 12, 316, 1945.

RYAN, D. M., "Penalty and Barrier Function," in *Numerical Methods for Constrained Optimization*, W. Murray (ed.). New York: Academic Press, 1972.

SCHOEFFLER, J. D., "Static Multilevel Systems," in *Optimization Methods for Large-Scale Systems with Applications*, D. A. Wismer (ed.). New York: McGraw-Hill, 1971.

SHAMIR, U., and C. D. D. HOWARD, "Water Distribution Systems Analysis," *Journal of the Hydraulics Division*, ASCE, vol. 94, no. HY1, 219, 1968, and vol. 95, no. HY1, 481, 1969.

SMITH, B. T., J. M. BOYLE, B. S. GRABOW, Y. IKEBE, V. C. KLEMA, and C. B. MOLER, *Matrix Eigensystem Routines-EISPACK Guide*. Berlin: Springer Verlag, 1974.

SMITH, C. L., R. W. PIKE, and P. W. MURRILL, *Formulation and Optimization of Mathemetical Models*. Scranton, Pa.: International Textbook Co., 1970.

SMITH, J. M., *Mathematical Modeling and Digital Simulation for Engineers and Scientists.* New York: Wiley, 1977.

STEWART, G. W., *Introduction to Matrix Computations.* New York: Academic Press, 1973.

TIKHONOV, A. N., and A. A. SAMARSKII, *Equations of Mathematical Physics,* International Series of Monographs on Pure and Applied Mathematics, vol. 39. New York: MacMillan, 1963.

TODD, D. K., *Ground Water Hydrology.* New York: Wiley, 1959.

VARGA, R. S., *Matrix Iterative Analysis.* Englewood Cliffs, N.J.: Prentice-Hall, 1962.

WIDLUND, O. B., "On the Use of Fast Methods for Separable Finite Difference Equations for the Solution to General Elliptic Problems," in *Sparse Matrices and Their Applications,* D. J. Rose and R. A. Willoughby (eds.). New York: Plenum Press, 1972.

WILKINSON, J. H., *The Algebraic Eigenvalue Problem.* Oxford: Clarendon Press, 1965.

WILLOUGHBY, R. H. (ed.), "Sparse Matrix Proceedings," IBM Res. Report RAI, Yorktown Heights, New York, 1968.

WISMER, D. A. (ed.), *Optimization Methods for Large Scale Systems.* New York: McGraw-Hill, 1971.

YOUNG, D. M., *Iterative Solution of Large Linear Systems.* New York: Academic Press, 1971.

YOUNG, D. M., "A Survey of Modern Numerical Analysis," *SIAM Review,* vol. 15, no. 2, 503, 1973.

ZIENKIEWICZ, D. C., *The Finite Element Method in Engineering Science.* New York: McGraw-Hill, 1971.

Chapter 3

Computer Implementation
of the Model

At this stage of the model-building process, we have completed the development of a solution technique for the previously defined mathematical modeling problem. We have also determined that the problem is theoretically solvable and that our solution technique is practical and will lead to a usable solution. We are ready for the computer implementation of this modeling solution. By "computer implementation" we mean software development including computer program design, development, checkout, and maintenance.

This chapter includes a comprehensive description of the software development process. It covers the development of the computer program from computing-problem definition through program operation and maintenance. The development of the program is treated rigorously using techniques being practiced today. If these techniques seem new, it is only because they offer a significant departure from the ad hoc techniques used by many programmers. They are based upon the results of research in computer science, on many years of good programmers' practice, and on the disciplines used in engineering. Experience gained over the past ten years shows that when these software engineering techniques are rigorously applied, the resulting computer programs are developed within project cost and schedule objectives and are easy to maintain and relatively free of the problems often reported by program users. We hope that our discussion will equip the reader (student or practitioner) with the tools he needs to implement his model. The basic steps of the method proposed for computer program design are illustrated using the mathematical model, developed in Section 2.3 for steady nonlinear flow in a pipe network. Figure 3.1 relates the discussion of Chapter 3 to the discussions presented in other chapters of this book.

3.1 INTRODUCTION TO THE SYSTEMATIC DEVELOPMENT OF COMPUTER PROGRAMS

The development of a computer program or computer implementation of a mathematical model is one of the major steps in the mathematical model-building process. Without a computer program, the modeling computations may be difficult or impossible to carry out. The following conditions are necessary for successful completion of the program development step:

1. Complete, precise and nonambiguous specification of all processes which will be carried out by the program (computational, data transfer, etc.).

2. Proven and tested correctness of the program with respect to these specifications.

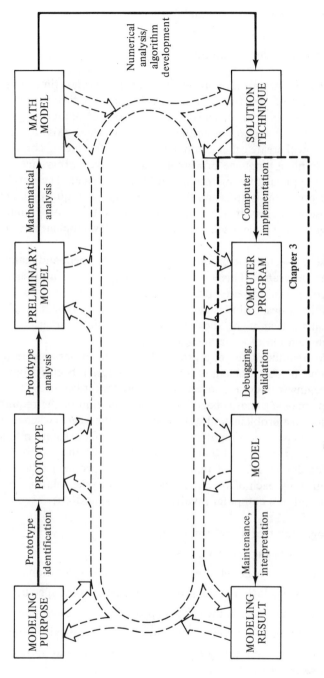

Figure 3.1 Guide to the discussion in Chapter 3.

3. Proper program structure and format.

4. Program coding in a suitable language.

5. Full program documentation.

Computer program development is expensive and time-consuming. Therefore it is mandatory that this stage of the mathematical model-building process be properly managed. The development of a computer program is an art that must follow certain scientific disciplines. There exist no computer programs that will create other computer programs (excluding compilers and other programming language processors that translate programs from one language into equivalent programs in another language). The development of computer programs is an exacting and difficult task, requiring much ingenuity of those who carry it out—the programmers. Good programmers are well organized: they have a scheme for program development and follow it rigorously.

We present a systematic procedure for the development and maintenance of computer programs. This procedure is based on recent research and practice over a period of years. It is now widely accepted and used by professionals in universities, in industry, and in government. It was developed to ensure the satisfaction of the above listed five conditions for success of computer implementation of a mathematical model.

It would be difficult to get a group of computer programmers to agree at which point the development of a computer program begins. Some would say that it begins with coding, others would say that it begins with program design, a few may say that it begins with the definition of the requirements of the solution to a problem, and some may say that it begins when a problem is perceived to exist. If a computer program is part of the solution to a problem, then the development of that program begins with the perception of the problem. The development of the solution to any problem follows a well-defined set of steps, each of which produces a result. The solution to a problem with the aid of a computer involves the following steps:

1. Recognition that a problem exists,

2. Definition of the problem,

3. Specification of the requirements of the problem solution,

4. Design of a solution to the problem that satisfies the specifications,

5. Adaptation of part of the solution procedure to the computer,

6. Construction and testing of the program as specified by the design,

7. Certification that the program is correct and that it satisfies the requirements specification,

8. Installation of the program in the computing environment in which the problem solution is carried out,

9. Maintenance of the computer program as a correct, efficient, flexible, economical, and useful part of the solution procedure.

We have addressed, in Chapters 1 and 2, most aspects of the first four steps. In fact, the model builder who has followed the steps outlined so far will have recognized that a problem exists and will have defined this problem. He will also have developed specifications for the problem solution and designed the solution process (algorithm). In our case then, discussion of computer program development begins with step (5). In general, however, the above nine steps are the major phases in the development of a computer program. In each of these steps a result is produced that is either used by another step, or is part of the developed program. Each step can be further broken down into tasks. The result of each task is part of the result of the step, and the collection of all these results is the computer program.

Our method of systematic computer program development identifies the results (or outputs) of the development project, the information or data (or inputs) from which these outputs are produced, and the tasks that produce the outputs from the inputs. These outputs, inputs, and tasks are specified from the most general to the most specific. Once specified, they are arranged in the order in which they should be performed. The process of program development, then, reduces to the performance of these tasks in the specified order.

Reviews of the process should be conducted at certain predetermined points, to determine that it is progressing properly. As a result of the review further progress, or revision of already accomplished tasks may be initiated.

In this introductory section each of the nine program development steps is described in summary; the output and the input of each step is described, the tasks that produce the outputs from the inputs are identified and discussed, and the sequences in which these tasks are performed are defined. The last five steps are covered in detail in later sections of this chapter.

3.1.1 The Recognition That a Problem Exists

The output of this step is a statement that things are not the way they should be and a list of symptoms that have led to that conclusion. The recognition of the symptoms of a problem or the perception that a problem exists do not constitute the definition of the problem. We must be very careful in this step not to confuse the symptoms of a problem with the problem itself. In many cases the recognition that a problem exists is a subjective process, but in other cases objective measurements can be taken to determine the exact nature of the problem and the seriousness of the problem. In both cases a model of the way things are and a model of the way things should be must exist or must be created. These models should be built as objectively as possible, then they should be reviewed by someone other than the person who created them.

In the case of interest to us here, that of mathematical model development and use, we have already recognized earlier that a problem exists. This has occurred when questions arose in "prototype space" for which answers were to be obtained by means of modeling (see Subsection 2.1.1). This problem recognition has lead to the initiation of model development.

3.1.2   The Definition of the Problem

During this step, the problem that has been perceived to exist must be defined completely, precisely, and unambiguously. The input to this step includes the symptoms of the problem and any additional information that will assist in the problem definition. The output of this step is the definition of the problem.

Problems develop when there is a difference between the way things are and the way things should be. Frequently, abstract models of the way things are and the way things should be must be constructed to detect problems. Once these models have been developed, their differences can be recognized. The task of problem definition is to develop these two models and then detect the differences.

It is easy to be careless in the definition of a problem, and programmers have not always pursued the definition of the problem until the right problem has been defined or until the problem has been well defined. In many cases large amounts of money, time, and other resources have been expended on the development of programs that solve the wrong problem.

In the case of interest to us here, that of mathematical model development and use, we have defined the problem earlier in the process. Problem definition was done in several stages: first a prototype and modeling purpose and a modeling problem were defined (see Subsection 2.1.1). Then, following model formulation a mathematical problem and a solution algorithm for it were defined (see Subsections 2.1.2–2.1.4). This solution algorithm, the problem data and certain conditions constitute the problem definition for computer program development.

3.1.3   The Specification of the Solution Requirements

The principal output of this step is a set of precisely and unambiguously defined requirements of the problem solution. These requirements are:

1. Basic requirements, including inputs, outputs, and processes.

2. Supplemental requirements, including efficiency, maintainability, ease of change, etc.

Development of these specifications proceeds in iterative fashion, from the general to the more specific level. Both basic and supplemental requirements are specified at each level. All specifications should be documented as they are developed. Documentation starts with an outline, identifying the actual tasks to be accomplished to produce a requirements definition. If the tasks correspond to the outline, the outline becomes a work breakdown structure for the performance of these tasks.

The top-level specification of problem solution requirements in the case of interest to us here, that of mathematical model development and use, has been accomplished prior to this stage. Specification includes the model input and output and computational process—i.e., an algorithm (see Subsections 2.1.1–2.1.4). Also several supplemental requirements were defined at that stage. However, the specification step will have to be repeated, at lower levels of detail, during computer implementation.

Requirements definition and analysis should always be reviewed and evaluated to ensure consistency with higher-level requirements and feasibility of implementation.

3.1.4 The Design of the Solution to the Problem That Satisfies the Specifications

The objective of this step is to design a feasible solution to the problem that will satisfy the specified basic and supplemental requirements. Design is done in two steps: preliminary design and detailed design.

During the preliminary design step, alternative approaches to satisfying the requirements are developed and analyzed. These alternatives are evaluated and one is selected for continuation into detailed design. The alternatives, their evaluations, and the recommendation are reviewed. The results of this review determine whether the selection is final, whether it should be changed, or additional alternatives should be considered. In the detailed design task, the selected alternative will be designed to a level of detail that is sufficient for program design and program construction.

The output of the detailed design task is a description of the solution to the problem independent of a computing environment. The solution procedure is represented using a complete and precise computational model. In our case this would include the solution algorithm, data, and miscellaneous specifications. This model depicts the structure of the solution, identifies every function within that structure, defines the input and output sets of every function and all of the operational sequences of those functions, and specifies the conditions under which each of the operational sequences will be executed.

This model becomes the basis for the design of the computer program. The design of the computer program is actually an adaptation of this model to a computing environment. The adaptation should be recorded in a design document, and used as a blueprint for program construction. There are no design problems remaining after this document is completed.

3.1.5   Adapting the Solution Procedure to the Computer

During this step, the abstract computational model produced in the previous step is adapted to the computing environment. The principal input to this phase is the abstract design document. The principal output is the program design document.

The abstract computational model depicts a solution to the problem. That solution has been developed relative to the problem itself. The program designer has intentionally kept the computer specifics in the background. Adapting an abstract solution to the computer should not be a major task. In some cases it is very simple, but in others it does require time, ingenuity, and a thorough knowledge of the computing systems on which the solution is to be implemented. During this phase, the program designer identifies the various blocks of code that will be developed as the main program, subroutines, functions, macros, etc. The interfaces between these program modules are defined. The data structures that will contain data are planned. The program design is optimized subject to the constraints specified in the requirements document and also to the computing environment. Program design should not alter the basic structure that was developed during the design of the abstract model.

3.1.6   Construction and Testing of the Program as Specified by the Design

In this step the computer program is coded and tested. The program design is used as a blueprint for coding. The symbolic test cases previously used to verify the correctness of the design are used as a basis for the actual program test cases. The construction plan that was formulated in the previous step is pursued here. The plan identified a minimum program configuration which is coded initially. Within the minimum program configuration, program modules are coded and tested in the order in which they can be used. This eliminates the requirement to store coded program modules until they are needed.

Program modules are coded in a language that is easy to read, easy to use, well suited to the process being coded, flexible, and available on each computing system on which the program will be executed. Programming language constraints are often imposed as part of the requirements specifications, and many installations impose programming language standards for certain kinds of programming. These constraints and standards should be followed; however, there are special cases when exceptions must be made. These situations can be determined best during the program design.

The program construction and testing step produced the program, test cases, and supporting documentation. At the completion of this step the tasks of the program development have been completed. The program and the supporting documentation are ready for acceptance testing and certification.

3.1.7 Certification That the Program is Correct and That it Satisfies the Requirements Specification

The computer program and its supporting material have been completed in the previous step. At this stage they are certified as being correct and meeting the specified requirements. It is desirable to subject these products to testing by an independent analyst or group of analysts. The purpose is to avoid the bias of the program developer(s). The program is tested for correctness and to ensure that is satisfies the requirements specifications. Also the support documentation is reviewed. Any problems uncovered during the review will have to be corrected and the corrections retested. Finally, the program and its supporting material reach the point of acceptability and the program becomes certified.

3.1.8 Installation of the Program in the Environment of the Problem Solution

The completed program and its supporting materials are installed in their operational environment. The program is installed on the computing system. The users are trained to use the program. Communication is established between the user and the development staff in order to provide training, consultation, and solutions to problems that may be discovered in the program or its supporting material.

3.1.9 Maintenance of the Computer Program as a Correct, Efficient, Flexible, Economical, and Useful Part of the Solution Procedure

To fulfill its purpose, a mathematical model is often used for a period of time that far exceeds the time required for its development. During this period of use, the program will probably undergo changes. Changes are made to correct errors, to provide a new capability, to take advantage of new solution procedures, or to transfer the program to a new computing environment. Changes to a program should be made using the same carefully controlled methods that were used to develop the program. Program maintenance could be called continued program configuration control and program development. Program maintenance must include the maintenance of the specification, the design, the program, the test cases, and all documentation. The effects of a change must be assessed and controlled throughout the program change project. Many programmers have made changes to a program that worked well only to find that something that worked before no longer works. These are side effects to changes. A complete history of the changes to a program must be kept. There are many programs running on computers today for which there is no change record, and until they have been completely redeveloped their users are at a disadvantage.

3.1.10 Summary

In this introductory section we described the development of a computer implementation, or a computer program, of a mathematical model as a set of nine steps. An overview was given of these steps and their expected results.

The development of any computer program, small or large, is complex and requires creativity. Yet it is essential that the program developer(s) follow a disciplined stagewise approach such as the one outlined. We have also found that documentation of these steps and reviews of the results are necessary for successful project completion.

In the remaining sections of this chapter we discuss in detail those steps that have not been covered in previous chapters.

3.2 COMPUTER PROGRAM DESIGN AND ADAPTATION TO A COMPUTING ENVIRONMENT

Computer program design starts with the previously defined solution requirements and solution algorithms. Also given at the outset is information such as the computational environment(s) in which the program will operate, the type, variability and frequency of computations required, data availability and format, the model user(s), etc. The design and adaptation process results in information and specifications sufficient for the construction, testing, installation, and maintenance of the computer program for our mathematical modeling computations. This information includes all the supporting documentation.

This section deals with the design methodology and process. We show how to produce a design and how to prove its correctness. The proposed design method proceeds top-down, from the general level to detail. This minimizes design errors and will, in turn, minimize program errors ("bugs"). The importance of program modularity is stressed, and the reader is shown how to maximize it. The process of design documentation is outlined, and we show how to produce, simultaneously with the design process, a readable design document, structured identically to the program. A brief discussion of design reviews follows. Then there is a subsection on the adaptation of the computer program design to the specific computing environment within which the program will operate. And finally, following a summary of the proposed method of computer program design, we develop a general design of a computer program for the model of a steady nonlinear flow in a pipe network. The reader will recall that this model was developed through the analytical stages in Section 2.3.

3.2.1 Preliminary Remarks

A computer requires a step-by-step program of instructions to direct it in the computational process that provides the solution to the mathematical modeling problem. Problems are defined and their solution requirements and algorithms are expressed in the working language of the people who perceive, define, and specify the problem and its solution requirements and algorithms. A problem definition and the requirements specifications are not a step-by-step set of instructions. The task of the designer is to formulate this stepwise solution process and express it in a language that can be translated into the machine language of the computer. Computing systems provide translators for languages such as FORTRAN, ALGOL, COBOL, etc., and they provide libraries of programs; however, they do not provide programs that create other programs. That is the task of the designer and coder. The designer must formulate a solution procedure satisfying the requirements. Then he must adapt it to the computing system using the languages it can process and the building block programs that are available from its libraries.

Designing a solution process to a problem can be a very complex task requiring long periods of time. The solution must be correct and it must satisfy the requirements specifications. We cannot wait until a program has been designed, coded, and tested to determine whether or not it will work and how well it satisfies its requirements. If it does not work or it does not satisfy its requirements then it will be costly and time-consuming to change, and there is a reluctance to make the changes that will produce a good program. A better approach is to develop a general, yet complete, precise, and unambiguous solution to the problem at the beginning of the design phase. This model can be tested, analyzed, and reviewed to ensure that it is correct and that it satisfies its requirements before proceeding to more detailed levels of design. The development of such a general model is neither time-consuming nor expensive; therefore, it is feasible to produce more than one such model.

Each model represents an alternative solution to the problem, and each can be subjected to analysis and review to ensure that the best solution is selected. Once an alternative is selected, it can be specified to a level of detail that is sufficient for program design and construction.

The objective of the design step is to produce a model that can be used as a guide or a blueprint for program construction. Before a program is constructed, its design must be completed; i.e., the design model must be formulated, tested, analyzed, reviewed, and documented. The design of the entire solution procedure does not have to be completed before program adaptation to the computing environment and construction can take place. Instead, we can complete the design of a minimum configuration of the problem solution, and then proceed with its adaptation and construction.

Nothing can be constructed before it is designed. Each block of code that is constructed should be built from a completed design. If that completed design has met its design criteria and if the program is carefully coded from the design document, then the chance for errors is very small. Programmers who design and code line by line usually spend many hours looking for errors in their programs, and it is well known that finding an error in a program is much more difficult, time-consuming, and expensive than finding errors in a design. One of the best ways to ensure that a design is correct and that a program corresponds exactly to its design document is to have someone else review the design and the program.

A computer program defines a process that produces output from input. It either is, or can be translated into, a set of instructions in the machine language of the computer on which it will be executed. To design the computer program we shall assume a standard set of tasks to be performed. These tasks are applicable to the design of all processes within the problem solution, from the most general process to the most detailed subprocess. The tasks simply produce a model of a process which will satisfy certain requirements specifications, general or detailed. The same types of tasks are applied to the design of all processes and subprocesses. That is, we consider the design problem, at all levels, to be the same type of problem and the same procedure is applied in each case. The same design scheme is applied to preliminary design and detailed design; therefore, the only distinction between preliminary design and detailed design is that preliminary design yields a general model and detailed design carries the general preliminary design model into a detailed model.

The design scheme described in this chapter shall be called *iterative multiple-level top-down design*, or *top-down design* (*TDD*). We always begin with specifications of the general, the basic, and the supplemental requirements. General requirements define background information; basic requirements define the input, output, and processing; and the supplemental requirements define the constraints on the solution. A general model of a process that satisfies the requirements is developed. This model does not contain much detail, but it is complete, precise, and unambiguous. It is designed to satisfy the most general basic and supplemental requirements. These requirements were defined at the top level in the (so-called) *requirements tree*; thus the general model is referred to as the top-level design model. Once this top-level model has been completed, that is, designed, tested, analyzed, reviewed, and documented, the designer proceeds to the next level of design detail. The top-level model contains subprocesses that compute partial results and these processes will have to be defined next. Their requirements specifications were defined as part of the design process for the top-level model. These requirements are again both basic requirements and supplemental requirements. The designer applies the same scheme to these subprocesses as he applied to the top-level or parent process. This procedure continues in subsequent levels until all of the processes and subprocesses are expressed in terms of elementary subprocesses or computer operations.

Proceeding from the top level to the bottom level or from the general to the specific we have, at all times, a complete model of the computer program. This model is the basis for tests, analyses, and reviews of the design. No other design scheme provides this advantage. A designer will never proceed from one design level to another until all of the design steps have been completed. An important design step is the test or proof of correctness. If the designer proceeded from one level of design detail to another without first verifying the correctness of the design model relative to its basic requirements, an error that existed in this model would be propagated into the next level and so on until it found its way into the program code. Furthermore, a poor design will not be detected until too late, if the models that are constructed at each level are not analyzed and reviewed. Finally, the documentation of each designed process must be completed as part of the design. The final document contains all of the information about the model, and it is the basis for testing, analysis, and review. Experience has shown that if the document is not produced immediately it is either never produced or it is not a good document.

The same set of design steps is applied to the design of each process beginning with the top-level process. Therefore we need only learn one design scheme. Its tasks can be learned by programmers and applied to small programs or to large projects. This results in standardization of design models, testing techniques, analysis techniques, documentation, and review procedures, and it makes project management much easier. This is not to suggest that our design scheme will eliminate the need for creativity on the part of programmers. There is nothing in this scheme or in any other scheme that eliminates the need for ingenuity on the part of the program designer. Instead this scheme provides a framework in which to work. It is a guide that helps organize the work of an individual or a programming team so that organizational details do not get in the way of creativity.

In the following subsections we will outline the steps in the design scheme and describe techniques for applying each of the steps. Alternative representations for the program model will be discussed. A documentation scheme and format will be presented. This yields the final documentation as a by-product of the design process with little or no extra effort. In fact, the design document will be completed by the time the design is completed.

3.2.2   The Steps in the Top-down Design Scheme

The only function of a computer program is to produce the elements of an output set from the elements of an input set. This is why one of the elements of the requirements definition is a set of basic requirements—input, output, and process. These three elements are depicted in Fig. 3.2.1.

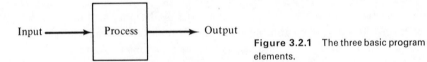

Figure 3.2.1   The three basic program elements.

The designer must specify the process that will transform the input into the output. The process may be subdivided into subprocesses and thereby specified. Each of these subprocesses has an input set and an output set and each transforms the elements of its input set into the elements of its output set. The subprocesses within the process must be executed in well-defined sequences. A particular sequence of subprocesses is selected or executed as a function of the input to the process. A model of the process must define these subprocesses, their input and output sets, the functional correspondences between the output set elements and the input set elements, the operational sequences of these subprocesses, and the criteria for selecting a particular operational sequence. This model must satisfy the basic requirements and the supplemental requirements. Every design scheme must identify or define these items and every design representation must depict the relationships between these items. The specific steps that are applied to the design of each process from the most general to the most detailed are:

1. Identify the subprocesses of a process,
2. Define the output set of each subprocess,
3. Define the input set of each subprocess,

4. Define the functional correspondence between elements of the output set and elements of the input set for each subprocess,

5. Define the operational sequences of the subprocesses,

6. Define the criteria for the execution of each operational sequence, and define the dependence of the subprocess output, within the operational sequence, on the subprocess input,

7. Prove that the design model (i.e., the results of steps (1) through (6)) for the process is correct,

8. Verify that the design model for the process satisfies the requirements specifications,

9. Document the design model,

10. Review the design,

11. Repeat from (1) for each subprocess that was defined, until further decomposition into subprocesses is impossible or undesirable.

In Subsections 3.2.3–3.2.11 each of these steps will be explained and illustrated. The final subsection (3.2.12) deals with the adaptation of the design to a computing environment.

3.2.3   The Subprocesses Within a Process

The objective of this step is to identify the subprocesses within a process. One of the questions often asked is whether there exists an algorithm that will decompose a process into subprocesses. The answer to this question is not a simple yes or no. If the output set of a process has been decomposed into its first-level elements, a subprocess can be identified, corresponding to each element of the output set. The resulting decomposition will eventually lead to a set of subprocesses that will produce the output set from the input set. This set may not be optimal relative to the supplemental requirements; e.g., it may not be efficient. It may be a good idea to use this scheme to get the design process started. The resulting subprocesses can then be combined, decomposed, and recomposed in order to improve the design model.

Is there a single best decomposition of a process? Will every designer decompose a process into the same set of subprocesses? Assuming that every decomposition leads to a correct problem solution, the decomposition must be evaluated relative to satisfying the supplemental requirements of the decomposed process. There may be more than one decomposition at the first design level that will result in a correct algorithm and satisfy the supplemental requirements.

The criteria that one can use as guidelines for decomposition include the definition of the output set, the processes that have been previously coded and filed in libraries, the need to use existing processes, the future usefulness of the processes developed, the limitations on the size of the processes, and the limitations on the execution time. Other criteria will be discussed below, and the reader will develop additions to these as he practices this design scheme.

The decomposition of a process can be represented by a *design tree*. The root node of the tree represents the process and each identified subprocess is represented by a branch emanating from the root node. Let a process be denoted by the symbol A. Suppose that the decomposition results in four subprocesses. These are labeled A.A, A.B, A.C, and A.D. Note that each subprocess which is at level two in the design tree has a two-syllable name and that each name has the name of the parent process as a prefix. This structured name assists in locating the node representing a process in the design tree, in the design document, and in the program. The two-level design tree for the above example is shown in Fig. 3.2.2.

In any design problem the root node of the design tree represents the requirements specifications or "what" is to be done and the next level represents the subprocesses which are part of the "how" or the solution.

There are variations to this design tree. One is to draw the tree with the root node at the left and the level two nodes to its right as shown in Fig. 3.2.3. This has the advantage that the nodes read from the top to the bottom of the page on which the tree is drawn just as their corresponding code may be printed in the program listing.

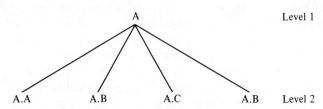

Figure 3.2.2   A two-level design tree.

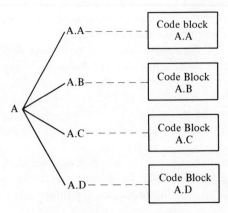

Figure 3.2.3   A design tree drawn corresponding to a program listing.

The left-to-right or the top-to-bottom arrangement of the nodes within a level in the tree is not arbitrary. These nodes represent the subprocesses of the process represented by the root node. Each of these subprocesses has an input set and an output set. The input sets of some subprocesses may contain elements of the output sets of other subprocesses within this design tree. In those cases where a subprocess requires the output from another subprocess as input, it must follow that subprocess in the operational sequence. In the tree it should be placed to the right or below the subprocess producing the output, depending upon the way the tree is placed on the page.

As an illustration, suppose we are to develop a computer program that computes the arithmetic mean, standard deviation, and confidence intervals for a set of numerical samples. The input to this program is the set of sample data and the output set will contain the mean, the standard deviation, and the confidence intervals.

The elements in the output set suggest that four processes are required at level two. These processes compute the arithmetic mean, the standard deviation, the first confidence interval, and the second confidence interval. The addition of an input process and an output process may complete this set of second level processes. Such a decomposition is an excellent way to begin the design. Output directed decomposition ensures that the elements of the output set are produced. The efficiency of the resulting design and other functions such as error detection can be considered next, but the production of the output is essential. A basis now exists for design analysis and review, even though the design has not been completed. The results of the decomposition may prevail or they may be modified.

The two-level design tree for this decomposition is shown in Fig. 3.2.4. There is not enough room in the tree to describe the process of each node at level two; therefore, the nodes in the tree are named with an hierarchical name and the functional name of each process is placed in a list.

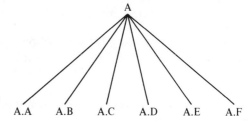

Figure 3.2.4 A two-level design tree for statistical computations.

The list containing the hierarchical and the functional names for the processes represented by the nodes in the design tree follows:

A.A   Read statistical samples,

A.B   Compute arithmetic mean,

A.C   Compute standard deviation,

A.D   Compute first confidence interval (mean $\pm$ standard deviation),

A.E   Compute second confidence interval (mean $\pm$ two standard deviations),

A.F   Print arithmetic mean, standard deviation, first confidence interval, and second confidence interval.

When the computer program is coded, each item in this list will appear as a comment at the beginning of an appropriate block of code. The processes represented by nodes A.D and A.E could be combined into a single process. If the process that reads the input checks the input elements according to some input specification, then it may recognize errors in the input data. This suggests a process that takes some action in case of errors, such as printing an error message and aborting the computations in the model. A library subprogram (see Subsection 2.1.5) may be available that computes arithmetic mean and standard deviation and makes both of these results available as output. This library program may be used for the combination of processes A.B and A.C. These types of considerations determine the final decomposition of a process into its subprocesses.

The decomposition task is the key and usually most difficult design task. It is not a mechanical task; but, rather, it requires the designer's knowledge. It is not done independently of the next two design steps: the identification of the output sets and the input sets of the subprocesses. This is additional evidence that design is an iterative activity—not only over the various levels of detail in a design but also within the design of the individual processes in that design model.

3.2.4 The Output Sets of the Subprocess

The objective of this step is to define the output set of each subprocess. The elements in each output set will be:

1. Used as input to other subprocesses within the same model,
2. Assigned to switch variables used to control the subprocess, and/or
3. Model output.

The definition of these output sets depends on process decomposition, input set definition, and/or operational sequence definition; i.e., the first four design steps are closely interrelated.

Each subprocess produces output or it would not be included in the design model. The elements of the output set of a subprocess are elements of the output set of its parent process and/or they are used within the model as input to other processes or as values assigned to switch variables. In the example used in the previous section, the output of the subprocess A.B is the arithmetic mean of a set of variables. This is used as input to the subprocess A.C, which computes the standard deviation, and it is also a member of the output set of the parent process, A. If A.A is designed to recognize input errors, then it will probably set a switch variable that is tested after A.A terminates to determine the next subprocess to be executed. Caution should be used when determining whether or not to use switch variables. A switch variable denotes the state of variables in a computation. It is usually better to let the state of those variables be tested outside of a process. In that way the program does not depend upon the setting of such variables within processes, and the process itself is independent of the program control logic. The process can be designed to satisfy its specifications independent of the containing program's control logic.

3.2.5 The Input Sets of the Subprocess

The objective of this step is to define the input set of each subprocess within the model. The designer chooses these inputs from the input set of the parent process based upon the functional decomposition of the parent process, and specifies the internal input requirements based upon the functional requirements of the subprocesses. The elements in each input set are:

1. Produced as output by other subprocesses within the same model, and/or

2. Equal to some of the elements of the input set of the parent process.

The elements of the input set of a subprocess are either equal to some of the input elements of the parent process or they are produced as output by other subprocesses within the model. In the latter case they are functions of the input of the parent process. Obviously, the input set elements of the first subprocess in the operational sequences within a model must be input elements of the parent process or, in the case that the subprocess is contained within a loop, these inputs will be outputs of the subprocess itself. The input to the statistical process illustrated above consisted of a set of variables. This set of variables is also the input to the subprocess that computes the arithmetic mean and it is part of the input of the subprocess that computes the standard deviation. The remainder of the input set to the standard deviation computation is the arithmetic mean which is the output of the previous subprocess.

3.2.6 The Functional Correspondence
Between Output and Input Sets

The elements of the output set of a process are produced from the elements of its input set. Each element of an output set is a function of one or more elements of the input set. This functional correspondence must be defined for each process. This is the objective of this design step. An output/input functional correspondence matrix can be used to record this information and it is an excellent display device. An output/input functional correspondence matrix is shown in Fig. 3.2.5.

INPUT SET ELEMENTS

Figure 3.2.5   An output/input functional correspondence matrix.

The matrix in Fig. 3.2.5 shows that O_1 is a function of I_2, I_4, and I_m. If the process that will produce an output from a set of input elements is known and named, that process name can be placed in the matrix instead of the ×'s. This matrix provides excellent visibility during design and design reviews. It also points out errors in specifications. Suppose that there were no marks in the column labeled I_4. This indicates a mistake in the specifications. Either input I_4 is extraneous or the defined functional correspondence for the output is incorrect. This matrix is also an aid to decomposition. A subprocess could be defined corresponding to each of the elements of the output set shown as row labels in the matrix. Later, we shall use this matrix in the proof that a design model is correct.

This matrix must be constructed for each process in the model, beginning with the top-level process which is defined during the requirements specification step. The output set and the input set of the problem solution should be defined at this time. This includes the identification of their elements and the description of the attributes of each element. The model builder should also determine the output/input functional correspondence at this time, and this definition becomes an important part of the requirements specification. During the design, the specifications for the subprocesses are defined, and this matrix is defined as part of those specifications. It will become a part of the design document for each subprocess. Later, during program maintenance, this matrix will be used to help assess the impact of proposed changes to the program.

3.2.7 The Operational Sequences of the Subprocesses

An operational sequence is the order in which the subprocesses are executed. There are a finite number of operational sequences for the subprocesses of a process and each operational sequence must be finite. It is unlikely, but possible, that a computer program will have only one operational sequence. In any case, the subprocesses in an operational sequence are executed one after the other in the order indicated, and this execution order is determined from certain criteria. It is the same whenever the criteria are the same.

The objective of this design step is to define and record the set of operational sequences of each of the subprocesses. The set must be recorded using readable notation that lends itself to testing and analysis, is flexible, and can be used as a model for program design and construction. Many notation methods can be used but not all satisfy these criteria. The set of all sequences could be listed, but this is not feasible if the number of sequences is large and the sequences are long. Furthermore, this notation does not lend itself to testing and analysis and it is not graphical. For example, the set of operational sequences of the four subprocesses in Fig. 3.2.2 represented in this way might be:

A.A, A.B, A.C, A.D;

A.A, A.B, A.C, A.C, A.C, . . . A.C, A.D;

.
.
.

A.A, A.B, A.D

etc.

It is obvious that this notation is not sufficiently compact. Furthermore, it is difficult to visualize the operational sequences represented by it. As we will see later, this representation, which is similar to the listing of a computer program is not convenient for testing, analysis, and review.

The directed graph notation satisfies most of the requirements. The vertices of the graph represent the subprocesses and the edges denote the order of execution of the subprocesses; thus they define all of the operational sequences. Figure 3.2.6 shows a directed graph for the four subprocesses A.A, A.B, A.C, and A.D. To this set of subprocesses has been added a begin and an end point.

The operational sequences can easily be read from this compact model by following the directed lines from the BEGIN point to the END point. Some of the sequences are BEGIN, A.A, END; BEGIN, A.A, A.B, A.D, END; BEGIN, A.A, A.B, A.C, A.D, END; and BEGIN, A.A, A.B, A.C,

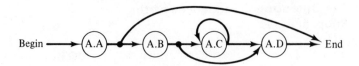

Figure 3.2.6 A directed-graph model of the operational sequences of four subprocesses.

A.C, A.D, END. Notice that the node A.C is contained within an arc which indicates that the subprocess it represents can be repeated. Notice further that each of the nodes has a single entry and a single exit. This implies that it is impossible to reach an interior entry point from outside of a subprocess and it is impossible to branch to another subprocess from within a subprocess. These subprocesses have been designed to meet a set of specifications that define their output, input, processing, and constraints. These specifications are independent of the specifications of other subprocesses. This must be the case in the design of systems; otherwise, a dependence exists between the operational characteristics of the processes. Suppose that subprocess A.B transferred to within the subprocess A.C. After the program had been in operation for awhile, A.C is changed. The designer of A.C, assuming that it was developed to satisfy its specifications only, did not make provisions for the transfer from A.B. When the new version of A.C is put into the program in place of the old version, the program no longer works. This happens frequently when existing computer programs are changed. The subprocesses should be developed to satisfy their specifications only.

A good rule to remember is the following: "A system should not have to look within its subsystems to operate properly, and subprocesses should not have to look outside in order to operate." This rule applies to both control paths and the transfer of data in and out of a subprocess. If this rule is followed in the design, the resulting program will have a modular structure.

In order to create flexible graphs that are readable, can be tested, can be analyzed, and that result in programs with simple structures, we limit the way in which the nodes can be interconnected. In other words, control graphs should be constructed from a certain set of subgraphs. This set of subgraphs includes a node with a single entry and exit, a node with a loop to itself, and a set of nodes only one of which is selected for execution. Each of these graph structures has a single entry point and a single exit point. Figure 3.2.7 illustrates each of these graph structures.

The broken lines drawn around each of these graph structures with single entry and single exit points illustrate that these structures can be interconnected only at single points. Böhm and Jacopini [1966] showed that a set of structures equivalent to these is sufficient to represent any computational process and therefore any computer program.

Program designs that are restricted to these simple structures (or an

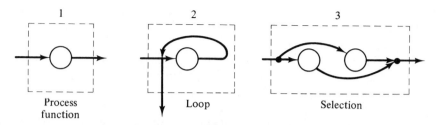

Figure 3.2.7 Three graph structures sufficient for representing sequential processes.

equivalent set) are simply connected and their operational sequences always flow in one direction (right or down, depending upon the orientation of the graph). The third structure that represents selection may include more than two nodes, only one of which is executed. All program designs illustrated in this chapter will use only these three basic structures.

A program design that is formulated using only the basic graph structures can be composed into a single node by substituting a node for each structure that is recognized. If this cannot be done the program is poorly structured and should probably be restructured. In Fig. 3.2.8 the graph is successively composed into a single node. This is equivalent to moving up the design tree from the bottom to the top. It is the inverse of process decomposition.

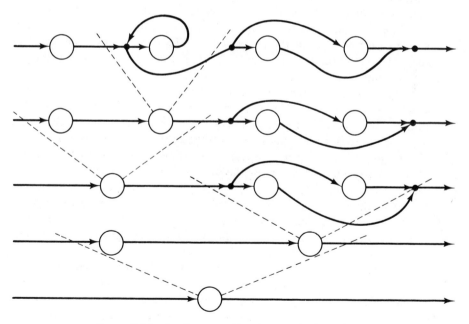

Figure 3.2.8 Graph composition by substitution of single nodes for basic structures.

　　　Designs for programs that are formulated using these basic structures are easier to read, test, analyze, and modify than those with unrestricted interconnections between modules. These three graph structures become an additional guideline to the designer during the formulation of the design model. The program can be coded from the design model using language elements that correspond to these structures or these structures can be constructed using the available language elements.

　　　The order of the nodes in a graph and the interconnections between the nodes can be established from the dependencies between the output and the input sets of the subprocesses that are represented by the nodes. If a process requires the output of another process as its input, then it must follow that process. In the statistical example, the process A.C, that computes the standard deviation is dependent upon the process A.B that computes the mean; therefore, node A.C must follow node A.B. If this output/input dependency does not exist then the ordering of the processes is arbitrary.

3.2.8　Criteria for Path Selection

In this design step each operational sequence or path through the graph is specified as a function of the input to the process. The input set may contain the assignment of a value to a switch variable (this variable is tested to determine the line to be traversed), it may contain the control variable for a loop, it may contain data that is used to compute a value that is tested to determine a path, etc. In each case the result of an input operation or a computation is a change in the state of the program variables, i.e., the program memory space. The selection of a path from a set of alternative paths requires examining the state of that memory space.

　　　There are many representations for program designs, including flow charts, pseudocode, decision tables, structure charts, state diagrams, etc. The notation that is used to represent the program design must be readable, must lend itself to testing and analysis, must be a suitable blueprint for coding, and it must present a complete, precise, and unambiguous specification of the program. No one of the representations used or mentioned satisfies all of these criteria well. Graphs are preferred as unambiguous representations for testing, analysis, completeness, and precision. They are especially useful at the higher design levels. Some people find them difficult to read at first. Pseudocode and structure charts are easy to read and make good program blueprints; however, they are not useful for testing and analysis.

　　　A state diagram may be used to represent an abstract sequential machine or process such as we are dealing with here. This diagram is a labeled directed graph, where the nodes represent states and the edges represent the transitions from state to state. Each edge is given a label, i/o, where i represents input and o represents output. Figure 3.2.9 shows part of a sequential machine.

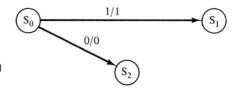

Figure 3.2.9 Part of a sequential
machine.

The machine depicted in Fig. 3.2.9 operates as follows: If the machine
is in state S_0 and a zero is read it transfers to state S_2 and outputs a zero.
If the machine reads a one while in state S_0 it transfers to state S_1 and outputs
a one. The change in state and the output imply that a next state function and
an output function exist. These functions can be expressed as

$$f(S_0, 0) = S_2$$
$$f(S_0, 1) = S_1$$
$$g(S_0, 0) = 0,$$

and

$$g(S_0, 1) = 1$$

The state diagram could be modified by placing the name of a process that
contains both the next state function and the output function in the edge as
a third component. Such a diagram is shown in Fig. 3.2.10.

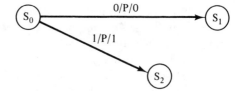

Figure 3.2.10 State diagram with
process names in edge labels.

Generally, this label takes the form input/process/output. A state diagram
labeled in this way suggests the exact relationship between a sequential
machine and the operational sequence diagram (or computer program).
A computer program consists of a set of processes that are executed sequentially. Each process transforms input into output. Each process produces a
new state of the program memory space. The diagram in Fig. 3.2.10 can be
inverted to produce the operational sequence diagram shown in Fig. 3.2.11.
The state of the program memory space prior to the execution of P_1 is S_0.
After the execution of P_1 the state of the program memory space is either
S_1 or S_2 as shown in both Fig. 3.2.10 and Fig. 3.2.11.

If the state after P_1 is S_1 then P_3 is executed. If it is S_2 then P_2 is executed.
The state change is a function of the present state and some input value. It
is the state of the memory space that is checked by the control statement in a
program to determine the process that is to be executed next. The state of the

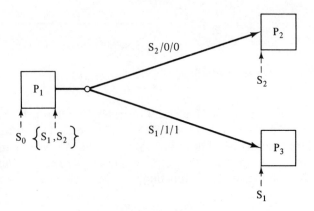

Figure 3.2.11 An operational sequence diagram showing state changes in the program memory space.

entire memory space could be examined. However, that is inefficient and usually also unnecessary. Instead, the designer specifies that only certain variables in the memory space have to be examined. These variables are expressed as conditions, such as $x \leq 37.5$, $i \leq n$, $y < x_1 \wedge y > x_2$, etc. These condition statements or predicates are used as labels on the edges of the graph that they cause to be traversed. The values of the variables within these conditions are computed by the processes as functions of their input; therefore, the unique input that produces these conditions is the second component in any label. The output of a process as a function of that input is the third component of every label.

Specifically, each line in the graph or operational sequence diagram is given a three-component label, c/i/o. Each line emanates from a process. The first component in the label represents the condition that causes the line to be traversed. The second component in the label represents the input to the process that causes that condition to be computed. The third component in the label represents the output produced by the process as a function of that input. Both the input and the output components represent vectors of one or more elements. The union of the input vectors in the labels on all lines emanating from a process is the input set of that process and the union of the output vectors is the output set of that process. The variables in the condition are also elements of the output set of the process.

The designer must define these labels for every line in the operational sequence diagram. The only exception is that the condition may be omitted from a label if the path is not one of a set of alternate paths from a process.

The labeled directed graphs shown are complete, precise, and unambiguous models of a program, an algorithm, a computation, etc. This model defines all processes and the sequences in which they operate. It defines the exact input that causes a path to be traversed and it shows the output produced along each path as a function of that input.

One can select a path through a graph. Following that path, the input in the labels may be recorded as they are encountered. Those input elements that are also elements of the input set of the parent process, are the inputs that cause the path to be traversed. The other inputs are internal to the model. This can be repeated for the outputs in the labels along the traversed path. The output elements that are members of the output set of the parent process make up the output vector of that path. As we shall see later, that output vector must be produced from the specified input vector of the path by the processes along the path. This will be used in the design testing procedure.

3.2.9 Testing and Design Model

At this stage of the model development we have identified subprocesses and their output and input sets. We have also defined the operational sequences of the subprocesses and the criteria for the execution of each operational sequence. This set of information constitutes a design model of a process. After the design model for a process has been formulated, it must be tested to ensure that it is correct. If the model is not tested and errors exist in the model those errors will be propagated into the next level design models and eventually into the program code. Diagnosing and correcting an error in a program requires more time and resources than the correctness proof of a design model. Figure 3.2.12 illustrates the potential impact of an undetected error in a design on the subprocesses in subsequent levels of design. That error has a potential impact on all of the descendants of the process that contains the design error.

An error left undetected and uncorrected in a design model can result in the requirement to redesign that model and many of its descendants, once the error has been diagnosed. The impact is even more severe after computer program construction, because the code must be modified or rewritten. There

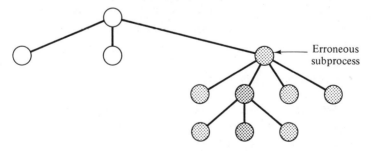

Figure 3.2.12 The fan-out impact of a design error on the descendants of the process containing the error.

is little justification for proceeding from one design level to another without first proving that the model is correct. Computer programs should never be coded until a design model has been completed and tested to ensure that it is correct. A design model should actually be proven correct analytically and the proof should be reviewed by technically competent people.

The first-level design model for each process should be tested immediately after it has been formulated rather than after the entire design has been completed. This not only eliminates the propagation of design errors but it also simplifies testing because the number of processes and the number of operational sequences in a model is smaller. Figure 3.2.13 shows an hierarchical model of a process and two levels of its subprocesses.

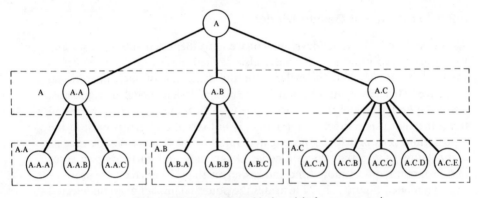

Figure 3.2.13 An hierarchical model of a process and its subprocesses.

In Fig. 3.2.14 the subprocesses within the model for process A and those within the models for A.A, A.B, and A.C are interconnected to show their operational sequences. Each of these models is simple, i.e., the number of processes is small and the number of paths through the graph is small. This is typical of the models created using top-down design.

In Fig. 3.2.15 the graph models for the processes A.A, A.B, and A.C have replaced the nodes A.A, A.B, and A.C in the graph model for A. As can be seen, the resulting model is more complex, containing more nodes and more paths.

One can imagine the complexity of the composite graph model for a design that contains ten levels. This is even worse when the undisciplined structure of many of the computer programs in use and under development is considered. For this reason the proof of correctness of many of the computer programs that are in use today would be an exercise in futility.

To verify that a design model is correct relative to the basic requirements, the designer must show that:

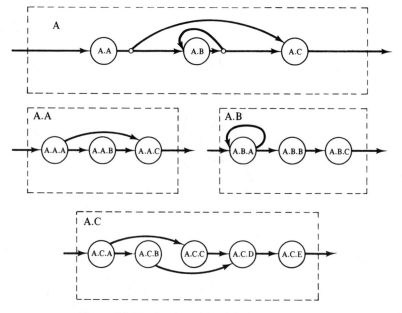

Figure 3.2.14 Graph models of the processes A, A.A, A.B, and A.C.

1. Every operational sequence of processes terminates, and

2. The elements of the output set can be produced from their respective input set elements.

The *first condition* requires that each process in the model be reachable from the beginning point, that the ending point be reachable from each process, and that transfer conditions exist that will actually cause each operational sequence to be executed. The *second condition* requires that exactly the elements in the output set of the process being tested can be produced. Each element in the output set must be produced from the input set elements specified. The output set elements must be produced by the processes within the model in the set of operational sequences depicted by the graph model. If any extraneous outputs are produced or if one or more output set elements are not produced then an error exists in the design model.

The first two parts of condition one can be tested by means of *connectivity analysis* of the graph model. While a careful visual examination of this model may reveal that it is properly connected, it is preferable to apply some algorithm to minimize the chance for human error.

To begin, construct a *connection matrix* from the graph. This square (Boolean) matrix will contain a row and a column for each node in the graph.

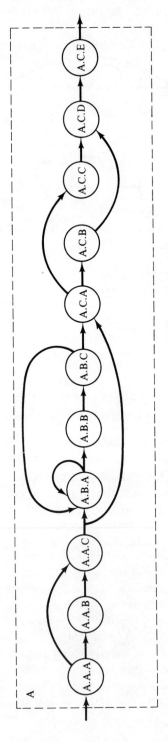

Figure 3.2.15 The composite graph model for three design models.

A one is placed in the connection matrix, C, at c_{ij} if the node represented by column j can be reached directly from the node represented by row i.

The empty connection matrix for the graph shown in Fig. 3.2.16 is shown in Fig. 3.2.17.

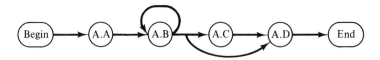

Figure 3.2.16 Example of graph.

	BEGIN	A.A	A.B	A.C	A.D	END
BEGIN						
A.A						
A.B						
A.C			c_{ij}			
A.D						
END						

Figure 3.2.17 An empty connection matrix for the graph in Fig. 3.2.16.

In the graph in Fig. 3.2.16 node BEGIN is connected directly to node A.A, but not the other nodes; therefore, the first row in the connection matrix will be (0, 1, 0, 0, 0, 0). Node A.B is connected to itself, to node A.C, and to node A.D. The third row in the connection matrix corresponding to node A.B is (0, 0, 1, 1, 1, 0). The connection matrix for the entire graph is shown in Fig. 3.2.18.

The connection matrix is actually a tabular representation of the graph. In this form it can be stored in the memory of a computer and it can be manipulated by a computer program. This matrix shows which nodes are directly connected to other nodes and it identifies all one-node loops by the presence of ones on the diagonal. If the graph has been constructed using only nodes in sequence, loops on single nodes, and/or selections, then the connection matrix will be upper triangular. A one in the lower triangular part of the matrix indicates that a cycle over more than one node exists in the graph or that the graph nodes are not in the best execution order. A loop containing more than one node can be reduced to a one-node loop by replac-

	BEGIN	A.A	A.B	A.C	A.D	END
BEGIN	0	1	0	0	0	0
A.A	0	0	1	0	0	0
A.B	0	0	1	1	1	0
A.C	0	0	0	0	1	0
A.D	0	0	0	0	0	1
END	0	0	0	0	0	0

— Loop over A.B
— Alternative paths

Figure 3.2.18 Connection matrix for the graph in Fig. 3.2.16.

ing the nodes within the loop with a single node, then letting these nodes reappear in the next design level.

After the connection matrix has been formulated, a reachability matrix is computed. The *reachability matrix* shows which nodes can be reached from other nodes by paths of length one, two, etc. Note in the connection matrix that a one exists in colum A.A of row BEGIN indicating that node BEGIN is connected directly to node A.A. Since any node reachable from A.A is also reachable from BEGIN, the particular logical combination (to be defined below) of rows BEGIN and A.A shows which nodes are reachable from BEGIN with paths of length two. Once this new row has been formed, a one can be found in column A.B indicating that all of the nodes reachable from A.B are also reachable from BEGIN; therefore, row A.B is logically combined with the new row BEGIN. This procedure is continued until every element of row BEGIN has been examined, then it is repeated for the remaining rows in the matrix. The result of this computation is the reachability matrix. The reachability matrix for the connection matrix shown earlier is shown in Fig. 3.2.19.

Each row in the reachability matrix represents the reachability vector for the graph node whose name appears on the row. The ones in the reachability vector tell which nodes can be reached from the node whose name is on the row. Each column in the reachability matrix represents the reaching vector

	BEGIN	A.A	A.B	A.C	A.D	END
BEGIN	0	1	1	1	1	1
A.A	0	0	1	1	1	1
A.B	0	0	1	1	1	1
A.C	0	0	0	0	1	1
A.D	0	0	0	0	0	1
END	0	0	0	0	0	0

Figure 3.2.19 Reachability matrix R computed from the connection matrix C (Fig. 3.2.18).

of the graph node whose name is on the column. The ones in this vector tell which nodes can reach the node whose name is on the column. The first row and the last column are especially important. The first row tells which nodes can be reached from the BEGIN node and the last column tells which nodes can reach the END node. The off-diagonal elements of these two vectors must be one. Suppose the BEGIN row in the reachability matrix had been (0, 1, 1, 0, 1, 1). The zero in the fourth position indicates that node A.C cannot be reached from the beginning of the program, and thus the design is in error.

The following algorithm produces a reachability matrix R of order n from a connection matrix of order n (Ramamorthy [1966]). A well-ordered graph containing no left-directed paths is assumed:

1. Set $i = 1$.
2. For all j, if $C(i, j)$ is true, then for $k = 1, 2, \ldots, n$ set

$$R(i, k) = C(i, k) \lor C(j, k)$$

3. Set $i = i + 1$.
4. If $i \leq n$ repeat from (2); otherwise stop.

This can be expressed in ALGOL as:

```
comment   Compute a reachability matrix of order n from a connection
          matrix of order n for a well-structured graph;
begin     Boolean array   C,R[1: n, 1: n];
          integer   i, j, k, n;
          for i := 1 step 1 until n do
                  for j := 1 step 1 until n do
                          if C[i, j] then
                                  for k := 1 step 1 until n do
                                          R[i, k] := C[i, k] ∨ C[j, k]
end
```

This program will not compute the reachability matrix for an arbitrary directed graph which may contain loops over many nodes or left-directed paths. The graph in Fig. 3.2.20 is not well ordered for it contains a left-directed path. The connection matrix for this graph and the reachability matrix computed by the program is shown in Fig. 3.2.21.

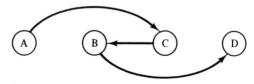

Figure 3.2.20 A poorly ordered graph.

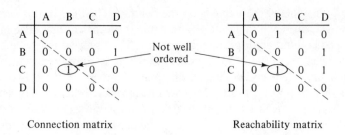

	A	B	C	D
A	0	0	1	0
B	0	0	0	1
C	0	1	0	0
D	0	0	0	0

Not well ordered

	A	B	C	D
A	0	1	1	0
B	0	0	0	1
C	0	1	0	1
D	0	0	0	0

Connection matrix   Reachability matrix

Figure 3.2.21 The connection matrix and an incomplete reachability matrix for a poorly ordered graph.

Notice the one in the lower triangular part of these matrices which indicates a left-directed path. Since this one is not contained in a square matrix along the diagonal whose elements are all one, it does not indicate a node within a loop of one or more nodes. Column D in row A of the reachability matrix contains a zero; however, an examination of the graph shows that node D is reachable from node A. The problem is that a one is introduced to the left of column C but the program does not go back and check for these new ones. An algorithm developed by S. Warshall (see Warshall [1962]) computes the reachability matrix for an arbitrary connection matrix. The rows are combined together as before, but the testing is done to recognize the left-directed paths immediately.

1. Set $i = 1$.
2. For all j, if $C(i, j)$ is true, then for $k = 1, 2, \ldots, n$ set

$$C(j, k) = C(j, k) \lor C(i, k)$$

3. Set $i = i + 1$.
4. If $i \leq n$ repeat from (2); otherwise stop.

This can be expressed in ALGOL as:

comment Compute a reachability matrix of order n from a connection matrix of order n for an arbitrary directed graph. The reachability matrix replaces the connection matrix;

begin   **Boolean array** $C[1:n, 1:n]$;

         integer i, j, k;

         for $i := 1$ **step** 1 **until** n **do**

              for $j := 1$ **step** 1 **until** n **do**

                  if $C[j, i]$ **then**

                      for $k := 1$ **step** 1 **until** n **do**

                         $C[j, k] := C[j, k] \lor C[i, k]$

end

The reachability matrix computed from the connection matrix in Fig. 3.2.21 using Warshall's algorithm, is shown in Fig. 3.2.22. It shows that there are no loops in the graph, that one left-directed path exists, that all nodes can reach D, that all nodes can be reached from A, and that node B cannot reach node C.

Figure 3.2.22 The reachability matrix for the connection matrix shown in Fig. 3.2.21.

	A	B	C	D
A	0	1	1	1
B	0	0	0	1
C	0	1	0	1
D	0	0	0	0

Many computer programs are constructed with left-directed paths (control that jumps upward in the program listing). This may indicate an error. It at least makes the program more difficult to read, to modify, etc. This has led many people to suggest that the GO TO statement should not be used in a computer program. This is an unnecessary restriction. The program should be carefully designed, using graphs and connectivity analysis as aids, then the code should be written directly from the design model. The resulting code may contain GO TO statements used to construct the flow of control that the designer intended. Programs coded in FORTRAN and in assembly languages must use the GO TO statement or its equivalent (transfers, jumps, etc.). An excellent paper on the use of the GO TO statement has been written by Knuth [1976].

Graph connectivity analysis is an excellent tool that can be applied to program designs and to programs. The program to compute the reachability matrix is simple and the program that will analyze this matrix is also simple. These programs should be developed and used as design and program analysis tools. When applied to an existing program, they can identify problems of the type illustrated that may otherwise go undetected or that have not been detected. Another extension to these programs would read the source program (FORTRAN, ALGOL, PL/1, COBOL, etc.) and develop the directed graph in tabular form, i.e., the connection matrix. Automating the formulation of the connection matrix, the computation of the reachability matrix, and the analysis of the reachability matrix reduces the chance for human error and increases the reliability of the program design and thus the program.

Graph connectivity analysis can be applied to a graph constructed from an existing computer program. The following steps will yield this graph:

1. Mark all control statements.

2. Identify the blocks of code between the control statements, give each block a label, and represent each block with a graph node.

3. Draw the directed lines in the graph as indicated by the marked control statements.

This type of analysis can identify many of the problems in a computer program. It will identify all upward control paths, all loops, the blocks of code that cannot be reached from the entry point and that cannot reach the exit point, as well as which nodes can be reached from other nodes. Also, if the reachability matrix is computed using addition instead of logical "or" the number of alternative paths from one node to the other nodes is counted. Graph connectivity analysis identifies only structural problems with a program. Many other problems may exist that will have to be identified using other techniques.

The reader will recall that we are dealing with verification of the design model correctness. Two necessary conditions were stated. The first is that every operational sequence of processes must terminate. The second part of the first condition requires that each *directed line* in the graph model can actually be *traversed*. A condition must be included in the label on each line and each of these conditions must be computable from the input elements in the same label.

Suppose that a node in a graph model represents a process that reads a set of numbers to be used in a computation. After the input process has been completed, the results are checked to determine whether the data read are correct or not. One of the paths following the input process is labeled with the condition "input correct," and the other path is labeled with the condition "input error." It must be assumed that a correct subprogram for the input operation will be developed. This means that the input data will be read and a variable will be set, that tells whether the input data are correct or not. We do not care how the input function operates, but only that it satisfies its specifications. The validity of the conditions in the labels then depends only upon the specifications. This reduces the verification of part two of condition one to checking the graph labels for valid conditions. A condition is valid if it is part of the specification for a subprocess that is executed before the directed line whose label contains the condition is encountered. It is usually in the label on the subprocess from which the line emanates. In the example, the specifications for the subprocess must include an element in the output set that indicates success or failure. If an input subprogram is used that does not include this element in its output set, then the error condition cannot be set and that path will not be traversed, thus the program may operate with incorrect data. In this case the "proof-of-correctness" of those paths that require correct data is not valid. Unfortunately, a large number of computer programs in operation that expect correct data do not check that data.

A special case must be made for each loop in a path. A loop acts like a function. It has a specific input set and it produces a specific output. In addition there is a termination condition for a loop. Each loop in a program consists of initialization of the loop control variable, testing the loop control variable, executing the statements in the body of the loop, and modifying the loop control variable. The loop control variable may be a specific variable independent of the variables in the loop itself or it may be one of the output variables of the loop. Test sets must be selected that will cause the loop termination to be tested and to produce the output of the loop. At least two specific termination conditions should be tested. An input set should be selected that will cause the statements in the body of the loop to be bypassed and another set should be selected that will cause the statements in the body of the loop to be executed at least once.

The second condition for design correctness requires that the elements of the *output* set can be produced from specified elements of the *input* set. The output/input dependencies have been specified and may have been displayed in a matrix such as that shown in Fig. 3.2.5. It is this specification that must be satisfied.

The output/input specification (output predicate) can only be satisfied by the operational sequences of processes defined along the paths in the graph. Verifying that this specification is satisfied requires that every path be traversed and that the input and output vector of each path be recorded from the graph labels as they are read. After recording the output vector and the input vector for a path, the elements that represent internal output and input elements must be removed from these vectors. The vectors that remain for the path must be precisely those in the specification. The set of all output vectors and their respective input vectors obtained by traversing all paths in the graph must equal the output/input specification.

The design verification procedure shown is based on symbolic representation of the conditions, inputs, and outputs, and it assumes that the processes along the paths in the graph are correct. In a sense it is a symbolic interpretation or execution of the design model (King [1976]). It produces, as a byproduct, symbolic test sets. These test sets should be recorded in the design document. At program test time, actual values can be assigned to each symbol. Programs cannot be exhaustively tested, yet test sets must be selected that will test the program paths. Symbolic interpretation of a design yields exhaustive testing, especially if the design proceeds top-down, because each model to be tested is relatively simple. Later, during program testing, values can be assigned to these symbols resulting in good test sets. Picking test sets for programs without such a basis is risky.

Design correctness is always verified relative to a requirements specification. At each level of design these specifications for a process are a product of a design. The requirements specifications for the program were a product of the design of the system that contains the program. If these specifications are not complete, precise, and unambiguous the design cannot be correct.

There is no way to prove that those specifications are correct; however, there are some guidelines for checking them. The reader will recall that the specifications for a process are derived from the design of the system that contains the process. Every process interfaces with other processes as shown in Fig. 3.2.23.

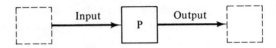

Figure 3.2.23 The interfaces of a process with other processes.

A process gets input from other processes, it produces output for other processes and is attached to a containing system (or parent process). It must provide output that meets certain standards, it must not take too much time, etc. These specifications can be checked by examining the structure of the containing system, and the input, output, and other requirements of the processes with which the process being designed interfaces. Every process or system is contained within another process or system. Unfortunately, many of these containing processes or systems are poorly specified, designed, constructed, tested, and maintained. This can result in poor requirements specifications for the processes contained within such systems.

An excellent formal discussion of "proof-of-correctness" is contained in Katz and Manna [1976].

3.2.10 Design Evaluation

At this stage the design model should be evaluated to determine how well it satisfies the supplemental requirements. This evaluation can be done using analytical techniques, simulation, benchmarking, and reviews. Any type of evaluation requires a model as a basis. A detailed evaluation may not be possible at the general design stage because information needed for precise computations may not be available. The usability of a design can be evaluated at the general design stage because the user interface specifications are defined very early in the design process. Detailed design data will eventually be available for analysis. The results of this analysis may show that the design will not satisfy the requirements. This, will necessitate a modification of the general design at this stage. In that case the impact of that modification will have to be assessed and handled from the level that was modified to the lowest level previously reached. This analysis and modification process is an additional illustration of the iterative nature of the design process.

Design analysis is used to determine the expected performance of the program, the amount of storage space that will be required in main memory and in auxiliary memory for programs and data, the quality of the results, the ease of use of the system, and the ease with which the system can be moved to a new computing environment, can be enhanced, or corrected. If the design models are not analyzed as they are formulated, then the results of poor design decisions could be carried into the operational system. These results will be recognized in the operational system; however, due to cost, time, lack of human and computing resources, etc., there may be a reluctance to make the changes that are necessary to improve the system. In some cases a program development project should be terminated during the early stages of design, when the design analysis shows that the requirements cannot be satisfied. There are many systems in operation producing poor results at high costs that would not have been completed if design analysis had shown that a good solution to the problem was not possible or feasible. Design analysis should be done as soon as possible, i.e., as soon as the design models and design data are available. Design analysis is an important step in the first-level design of each function within a computer program.

Performance analysis determines the response of the system, i.e., the *time* required to execute a program, the time required for a system to respond to a request entered at a terminal, etc. At high levels of design the data required for detailed timing analysis may not be available. In those cases estimates must be made of the execution times of subprocesses. A timing analysis depends upon a model of the program being designed and operational data about the computing system on which the program will operate. The graph model for the program being designed provides the basis for deriving a timing formula. The data about the computing system must be obtained from documents, modeling and simulation, benchmark testing, and personnel who work with those systems.

In Fig. 3.2.24 let $T_{A.A}$, $T_{A.B}$, and $T_{A.C}$ represent the time required to complete the execution of the respective processes. There is only one operational sequence of the three processes in Fig. 3.2.24, therefore the execution time for the parent process A, independent of the host computing system is represented by

$$T_A = T_{A.A} + T_{A.B} + T_{A.C}$$

Very little may be known about the processes A.A, A.B, and A.C. Unless these are existing programs with available performance data, an estimate of the upper and lower bound on the execution time of each must

Figure 3.2.24 A program graph with one operational sequence.

be made. At the more detailed levels of design this data will become available. It can be used in similar formulas to compute values for time variables. A bound on the time interval required for the execution of each process can be computed and then substituted from the bottom of the design tree towards the top to validate estimates or computer performance data. This may result in required changes to the design models at high levels in the design tree. If a design model at a high level is changed then all its branches must be reviewed for required changes.

If a loop is added to the model to repeat the execution of A.B the formula becomes

$$T_A = T_{A.A} + n(T_{A.B} + c) + T_{A.C}$$

where n represents the number of times the function A.B is executed and c represents the time required to execute the loop control statements. If the time to execute control statements is small, then the timing formula becomes

$$T_A > T_{A.A} + nT_{A.B} + T_{A.C}$$

Suppose that an alternative path is introduced into the model shown in Fig. 3.2.24. There are now two operational sequences of the subprocesses. A timing formula can be written for each path through the program. Each formula will include the time required to execute the control statements that test conditions and cause that path to be traversed. A modification to the earlier model with an alternative path is shown in Fig. 3.2.25. If P_1 represents the

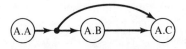

Figure 3.2.25 A program graph with two operational sequences.

probability of selecting path 1, then the timing formulas for the execution of each path in this model are:

$$T_A > T_{A.A} + T_{A.B} + T_{A.C}$$
$$T_A > T_{A.A} + T_{A.C}$$

The timing formula for execution of the total model is:

$$T_A > T_{A.A} + P_1 * T_{A.B} + T_{A.C} + c$$

where c represents the time to execute the control statements.

Timing formulas are easy to derive for graphs that are composed from the three basic graph structures. A timing formula is written for each path in the graph that includes processes in sequence and loops. A timing term is derived for processes and for loops. These terms are combined with the control statement timing to yield the formulas for each path.

The timing analysis shown does not consider systems in which many programs are competing for a single resource such as the central processor of the computer. Instead, it is intended for the analysis of programs as if they were to execute on a dedicated computer. The execution time bounds computed from these formulas are useful in comparing alternative design models; however, they do not yield the actual elapsed time that a program will require for execution in a computing system. That type of analysis requires information about the host computing system, the number and type of programs that are executing, etc.

The exact *storage* requirements for a program and data cannot be known until the program and the data have been compiled into the internal format of the computer storage. Estimates of storage requirements can be made at all levels of design and these estimates are improved with each new design level. It would be very difficult to estimate storage requirements before design begins. The only model that exists at that time is the requirements specification represented by the root node in the design tree. A better estimate can be made at level two because more design detail exists. At the most detailed level of design, actual storage requirements for existing modules can be used and good estimates can be made for the program modules that must be coded. Figure 3.2.26 depicts a design tree for a program. The nodes enclosed in squares denote existing blocks of code and the others represent

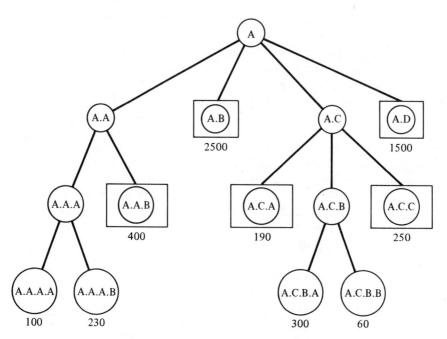

Figure 3.2.26 A design tree used for estimating program storage requirements.

program modules that must be coded. The actual and estimated storage requirements for the programs represented by these nodes are written beneath each node in the tree. These storage estimates and actuals are summed from the bottom of the design tree upwards yielding estimates and actuals for each subsystem and finally for the entire system. The results are shown in the following table.

PROCESS				ESTIMATED STORAGE (WORDS)			
A				5530			
	A.A				730		
		A.A.A				330	
			A.A.A.A				100
			A.A.A.B				230
		A.A.B				400	
	A.B				2500		
	A.C				800		
		A.C.A				190	
		A.C.B				360	
			A.C.B.A				300
			A.C.B.B				60
		A.C.C				250	
	A.D				1500		

If the storage estimate is required at level three before the remaining design levels have been completed, then estimates will have to be made for A.A.A and A.C.B instead of using actuals.

At the most detailed levels of design, storage required for each statement should be estimated. Average storage per statement can be used for this purpose.

A similar strategy can be used to estimate the storage requirements for program data. These estimates are usually easier to make and can be made more accurately because the data elements are well defined. For example, a program designed to work on an $N \times N$ matrix with real elements will have to allocate $W \times N^2$ storage elements. W represents the number of basic storage elements required for each element of the matrix. The total storage requirements for a program are the sum of the storage requirements for the code and the data.

An analysis of the *useability* of the program is best done by using the completed program; however, if the designer waited until the program was completed to get the user reaction to input formats, output formats, etc., it would be too late to make the changes that may be required to make the program easy to use. This type of analysis should be done as early as possible in the design process. It can be done by reviewing the user interface and by simulating the use of the program. The simulation can be done using paper and pencil or with a system that supports the creation and operation of user interface prototypes.

The user interface includes the input language, output language, and operational procedures of the program. User input and output languages are designed very early in the design process. An input language may consist of numbers to be punched on cards in a specific format, a command language for an interactive programming system, a sophisticated programming language, etc. The potential user of the program should be involved in the design of the user language, if only to review the language and practice its application. Output languages specify report contents and formats, the content and formats of displays on plotters, the layout of matrices, etc. The operational procedures for a program can also be specified early in the design process. As soon as the user interface has been specified, it should be made available to the user for practice and review. The user can fill out input forms, examine output layouts, etc., that are proposed by the program designer. In other cases designers have used prototype systems to allow the user to practice with a proposed user interface.

The evaluation of the user interface is very important, yet it is frequently ignored. It is the user who will use the program and not the programmer. The program should be designed for the user and its design should be reviewed by the user.

3.2.11 Design Documentation

The designed computational model begins as an abstraction in the mind of the designer. This model must be recorded in verbal, written and/or pictorial form. Such a record will make it possible to communicate the model to others, to perform analysis, to verify the correctness of the model, and to construct the computer program itself. The design document is the source reference for the design model. Developing this document should require little effort and should be done as the design model is formulated.

Documentation has always been an undesirable task for programmers, done after the program has been completed or not at all. Program documentation has frequently been difficult to read and understand, often completed late, and with little resemblance to the design or the program represented. In this section we show that the design document can be an immediate byproduct of the design process; it can be relatively easily comprehended, even by readers who are unfamiliar with the design. The document can and should be structured exactly as the design that it represents.

The design document should be modular. A design document section should be written for each module (i.e., process or subprocess) in the design model. These documents are composed as the design model is created, module by module, using a design methodology such as TDD. This ensures that the document is produced at the same time as the design model and that its structure is identical to the structure of the design model.

The document corresponding to a module in the design model is created by recording the steps in the design process as they are performed. The design document should contain exactly the information produced in the design process together with the specifications that are to be satisfied by the design. This information should always be recorded, and if it is recorded in a particular format, that record becomes the design document. This scheme produces the document in parallel with the design, it creates a document that is structured like the design model, and it results in a document that is readable from the general to the specific with the detail increasing level by level. In fact, the most general level documentation is completed first, and it is needed first because it has the most general readership. Since the document is modular, it is also relatively easy to maintain.

The steps in the top-down design process were defined in Subsection 3.2.2 as:

1. Identify the subprocesses of the process being designed,

2. Define the output sets of the subprocesses,

3. Define the input sets of the subprocesses,

4. Define the functional correspondence between elements of the output set and elements of the input set,

5. Define the operational sequences of the subprocesses,

6. Define the conditions for transition from subprocess to subprocess within the operational sequences as functions of the input to each subprocess; and define the output of each subprocess within the operational sequences as functions of the inputs to each subprocess,

7. Verify the correctness of the design relative to the basic requirements,

8. Analyze the quality of the design relative to the supplemental requirements,

9. Document the design,

10. Review the design.

The documentation of the design model is a recording of the results of steps (1) through (8) of the design process together with a summary and either the requirements specifications for the model or a reference to those specifications or both. These are the minimum contents of the document. The following format for a document is suggested for the purposes of discussion:

1. Design module (i.e., process or subprocess) name.
   (a) Hierarchical name.
   (b) Functional name.
2. Summary.
   (a) Function of the module.
   (b) Description of the method of solution.
   (c) Justification of the method of solution.
   (d) Evaluation of the model.
3. Requirements specifications.
   (a) Interfaces.
   (b) Output set.
   (c) Input set.
   (d) Functions.
   (e) Constraints.
4. List of subprocesses.
5. Graphical model.
6. Design test.
   (a) Test sets.
   (b) Expected test results.
   (c) Test procedure.
   (d) Test results.
   (e) Test evaluation.
7. Design analysis.

The *design module name* consists of two parts. The first part is the *hierarchical* or structured name of the module. This name indicates where the module resides in the design tree. It is used as an index into the design document and later into the program. The second part gives a short description of the *process*. An example of such a two-part module name is:

A.A.B.E: COMPUTE THE ARITHMETIC MEAN AND THE STANDARD DEVIATION.

The structured name, A.A.B.E, tells us that the module is at design level four and that its parent (or containing) module is A.A.B. The functional name gives a short description of the process, i.e., the module computes the arithmetic mean of a set of numbers and also computes the standard deviation of that set of numbers. The naming convention traditionally used by programmers is to create an abbreviation for the process. In the above case a programmer might use the name STDEV. Such an abbreviation provides no structural information and does not always provide a good indication of the process. To indicate the position in the tree of a common process which is used more than once, the location of the reference to that function should be given the structured name and the process should be given another name.

The *summary* is intended to give a brief description of what the module does, how it does it, and how well it does it. It is occasionally the only part of the document that will be read. It can and should contain tables, drawings, etc. It is a good idea to write this summary based on the other parts of the document to ensure its completeness and accuracy.

The *requirements specification* or a reference to the specifications gives the reader a list of the requirements to be satisfied by this module. This section also points to papers, books, and documents pertinent to the solution.

The *list of subprocesses* gives the first-level decomposition of the process being described. Each line in the list contains the structured name and the functional name of a subprocess. The list is ordered to show the functional dependencies of the items in the list upon each other. The items in this list correspond exactly to the end nodes in the design tree for the node represented by the design document.

The *graphical representation* is a precise pictorial-symbolical model of the process. It must allow the reader to produce the elements of the output set from the elements of the input set by symbolically executing this process. We recommend to use as models labeled directed graphs. These lend themselves well to review, analysis, proof of correctness, and reading. In some cases one may want to supplement the graph with other representations and in other cases there may be representations that can be substituted for the graph. Within a design document the expression of the model should be consistent; therefore, one should not use different models for the various processes except in those cases where a model has some definite advantage over another. Even in those cases one should adopt a standard which is used whenever possible.

The design *test* section lists the test input sets, the expected test *results*, the test *procedure*, the actual test results, and an evaluation of the test. The design test should be included in the document so that it can be checked by those who review the design. It is also used as the basis for the test of the program that is constructed from this design. Whenever the program is modified, the affected modules will have to be retested. This test section is the basis from which new test sets, procedures, and expected results are formulated.

The program *analysis* section describes the evaluation of the design model relative to the supplemental requirements, including an estimate of the expected program execution time, storage estimates, the expected accuracy of the results, an estimate of the volume of output, etc.

After a design has been formulated and the document has been written, the design is subjected to a technical review. The information contained in the document is the basis for this review. When the review is completed, the design document should be placed under "configuration control." The term *configuration control* refers to the management process which ensures that program changes are properly controlled and coordinated (see Section 3.4 for more on this subject).

Figure 3.2.27 shows a partial design tree and the corresponding document tree. One can see that they are similarly structured and that the structured name of the components in the design tree are used as the indices into the design document. The program listing can be considered an extension of the design document.

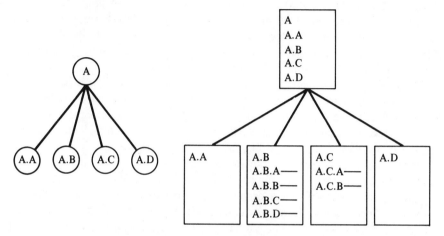

Figure 3.2.27 A partial design tree and corresponding document tree.

The title or heading of the (listed) program itself should contain the same name pair (structured and functional). A short summary should, follow this title as comments in the program listing. Next, the output set and input set descriptions should appear in the program listing. Each line in the list of subprocesses should be placed into the program as a comment and be followed by the actual code that performs the indicated process.

If the code is removed from such a program listing, the comments that remain should be (almost) identical to the document. This will establish a direct correspondence between the program and that document.

3.2.12 Design Reviews

The basic reasons for design reviews are to ensure that the design performs its function, that performance conditions are satisfied, and to solicit new viewpoints on a design or a program. A programmer working alone should always submit a design or a program to another programmer for review. On large-scale projects reviews may be conducted before examining committees.

In this chapter we have subdivided the program development process into phases. Each of these phases produces some result, and that result should be submitted to others for review. The result may be the requirements specification document. The review in that case would consist of examining the document to ensure that it is complete, precise, well structured, readable, and maintainable. To assist the review team in being consistent, a review checklist should be used to check the document. Additionally, the schedule, the budget, problems encountered, etc., may be discussed at such a review.

A design review should be held at the end of the design phase; however, during the design, reviews should be conducted for each completed design element. For certain critical modules in the program the design may have to be submitted to a review team. Other modules, on the other hand, may be reviewed by another designer. Every design in the program should be reviewed whether it is large or small, significant or insignificant.

A design is completed to satisfy a set of specifications. The process developed must transform the input into the output, and it must do it within a specified time, to a certain format, etc. A design review must determine if those specifications have been satisfied. In addition it should determine the quality of the work, the generality and flexibility of the design, it should judge whether the design satisfies the standards established for the project, etc. The source of materials for the design review is the design document and the requirements specification. A design review checklist identifies the items that should appear in the design document. The document standard shows the format for the document. The content and the format of the design document for a module can be judged objectively using the checklist and the standard. The quality, generality, flexibility, and other attributes of the design may have to be judged subjectively and through the documented analysis and the proof of correctness in the design document. The description of the design document in the previous section makes an excellent checklist for reviewing the contents of the design document.

3.2.13 Adapting the Design to a Computing Environment

The result of the design process described so far is a functional model, or a process representing part of the computer implementation of our mathematical model. Before this model can be coded as a computer program for a particular computing environment it must be adapted to that environment; the program structures and the data structures must be specified, interfaces between the program and other elements of the mathematical model must be specified, a program construction plan and a program test plan must be formulated.

For simplicity these elements were excluded from the description of the design process. In actual practice a designer would find it very difficult to ignore the elements of the host computing environment, such as compilers, programming languages, subprogram libraries, operating systems, hardware, etc.; but the program may be developed for more than one computing system or it may later be transferred to another computing system. Therefore, the designer should aim for a high degree of program portability. This requires that the functional model be as free of the constraints of a particular computing environment as possible. Again for simplicity of explanation the adaptation of the functional model to the host computing environment, the specification of the program and data structures, the interfaces, and the development of the program construction plan and the test plan will be described as if they were done independently of the functional model. In practice these activities may be parallel with design or they may be done after the design has been completed.

A computing environment consists of computing hardware, an operating system, and the standards and procedures under which the computing installation is operated. These components of the computing environment vary from one computing installation to another. The functional model must be adapted to conform to each of these in each computing environment in which the computer program will be used.

Computer hardware consists of one or more computers which may be interconnected. Each computer includes a central processor, memory devices, and input/output devices. The central processor performs operations on data. These operations include arithmetic, character manipulation, memory access, input, output, etc. Each operation requires time for its execution and this time varies on different computers. These operations may be done sequentially, in overlapped fashion, or in parallel depending upon the type of computer that is used. The results of these operations can vary on different computers. The memory of the computer is used to store the programs and the data. The memory includes a central memory and auxiliary memory. The instructions executed by the computer and the data processed by those instructions must reside in the central memory. Because the central memory of a computer is limited, the program and data may have to be placed in both the central memory and the auxiliary memory. If a program and the data must be stored in both memories the program must be divided into blocks of instructions and a program must be available to read the required blocks of instructions from the auxiliary memory into the central memory. Such a program is usually available as a utility within the operating system and in some cases a program can be segmented and transferred between the auxiliary memory and the central memory as needed automatically by the computing system.

The need for adaptation

The functional model as described in this chapter is an ideal model of a solution. The model consists of operands and operators. The operands represent the data that is to be processed and the operators represent the processes. The data may be integers, real numbers, complex numbers, character strings, vectors, matrices, records, files, etc. The operations that are performed are input, output, sorting, arithmetic, logical, character manipulation, etc. When creating the functional model, the designer may choose to remain unconstrained, e.g., he assumes that the operators will require no time for their execution, the results of the operations are error free, the axioms of a mathematical system are valid, data storage is infinite, etc. This approach to the development of the functional model simplifies the design process. If the designer constrains the design process (by minimizing the number of operations that are performed for example) he may still choose to ignore the constraints of the computing environment. This may not always be a good idea because the features of a particular computing environment may have a direct impact on the model; therefore, the use of these special features must be weighed against the impact of the loss of program portability.

The adaptation of the functional design to a particular computing environment must consider the differences between the computational environment and the mathematics applied during design, the limited storage of the computing system, and the time required to perform the computations. In addition the interfaces between the user, the data, the program, and the computing system must be specified. Also, the design tree must be converted into a program structure consisting of modules. The program modules, their interfaces and the data structures that will be constructed must be specified. The standards of an organization must be applied. After these things have been done, the program design is complete and ready for review. Once the approval of the program design has been obtained, the program, data, and job control language can be coded. Once all of the described steps have been completed satisfactorily, coding should be little more than transcribing the program design using the chosen programming languages.

The adaptation of the functional model to a computing environment is a very important part of the program development process. It must be done for each unique computing environment on which the program is to operate. It would be difficult to produce a computer program that could be transported from one computing system to another without adaptation because each is different. The number of digits that can be carried in a number vary, the arithmetic operations do not always produce the same result, the time required to perform operations varies, etc. Even though a program, as coded in a higher-order language, may execute similarly on various computers, the user interface to each computing system is different. To be certain of a correct implementation of the program, the mathematical model must be adapted to each unique computing environment on which it is to operate.

Adaptation is another approximation

A mathematical model is an approximation of a prototype, using exact integers and real numbers. A computer limits the number of digits in an integer or a real number. Special subprograms and extended precision arithmetic operations on computers may extend the number of digits that can be carried, but there is always a limit. This means that another approximation has to be made. The subject of finite precision floating-point arithmetic is discussed in more detail in Sections 2.1 and 4.4.

Execution time

The time required to execute a series of operations on a computer may be significant, it is proportional to the number of operations performed. The time required to perform an addition, subtraction, multiplication, or division may vary from one computer to another. The execution time for operations may constrain the model builder in using certain operations: for example floating point division on certain computers. Some computers overlap operations, i.e., they perform certain sets of operations in parallel. To take advantage of these features, special algorithms have been designed.

Storage of programs and data

The physical storage for programs and data in a computing system is limited. The storage usually consists of main memory and various types of auxiliary storage. The time required to access the main memory is relatively short; however, its capacity is usually small. The auxiliary memory devices, such as magnetic drum, magnetic disk, magnetic tape, etc., have a much larger capacity, but their access times are also greater. There is often not enough main memory to store a program and its associated data. A matrix of order 1000 would require one million words of main memory. On an IBM 370 this would require four million bytes of main memory if the matrix elements represented real numbers in short precision. This is more main memory than most of these computers have available. On smaller computers the available storage may be even more limited. The layout of the data structures must be such that they will hold the data. These layouts may vary from computer to computer, and they will affect the design of the computer program. Unfortunately, there are no computing systems available that do not require careful planning of the data structures used with a program and the allocation of those data structures to various types of storage devices.

Computer programs also require storage. When the program executes on the computer, it must reside in the main memory of the computer. Some computing systems segment the program into physical units and store only the active part of the program in the main memory while the inactive part is stored on auxiliary storage devices. If the programs fit into the main computer memory or if the computing system segments the program and allocates main memory for the active part of the computer program, the planning of the program structure is simplified. If this is not the case then the program must be segmented and special instructions must be coded that will read the required parts of the program from auxiliary storage devices into main memory.

Program segmentation

Program segmentation and the loading of program parts into main memory for execution require time. If the segmentation is done by the computing system then computer time is required for the segmentation. In either case the program segments must be read from the auxiliary storage devices and this reading operation requires time. This time must be considered by the model builder in his attempt to satisfy the performance requirements of the computer program.

The design tree depicts the processes of the model and the hierarchical relationships between them. These processes are shown from the most general (root node) to the most detailed (leaves). This tree is seldom developed to a level of detail that shows individual operations, such as arithmetic operations; however, that could be done. The design tree is the basis for planning the segmentation of the program. The program could be constructed with no segmentation, i.e., with the program modules (processes or subprocesses) coded completely in-line.

The design tree identifies every process in the model. Each of these may be implemented in-line or as a subprogram. The criteria used to select the subprograms include the availability of subprograms in libraries, the future utility of a process as a subprogram in other programs, the optimization of programmer time and storage, and the physical limitations on the size of a block of code. Many excellent subprograms exist in computer libraries at each computer installation, at other installations, or in libraries maintained by groups of computer users (see Subsection 2.1.5). In addition, many excellent mathematical models coded in publication languages, such as ALGOL-60, are available in journals and catalogs. These subprograms and algorithms should be used whenever possible. In fact, it is desirable to compose a program completely from existing blocks of code. The existence of these subprograms and algorithms will influence program design and its adaptation to a computing environment. The designer must examine the design tree for processes for which subprograms are available. When such a process is identified in the tree, it should be marked with a special symbol and a reference should be made at that point in the tree to the name of the library subprogram. The branches of that node in the process can be pruned from the tree. Figure 3.2.28 shows a design tree before and after a subprogram has been identified for a process. The asterisk marks the node as a subprogram reference and the reference in parentheses is to a list of subprograms.

The program designer may recognize many identical processes in the design model. Whenever the specifications for part of a program are identical to those of another part, the designer must consider implementing those program parts as subprograms. These can either be coded in-line or as a subprogram, referenced from each point as required. The decision to code the subprogram in-line or out-of-line is based upon the future utility of the

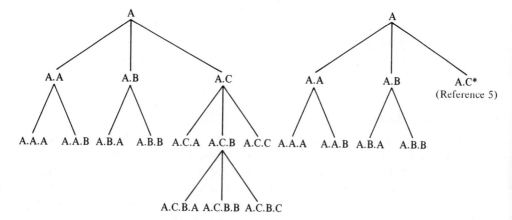

Figure 3.2.28 A design tree and the corresponding program tree.

subprogram, its physical size and the cost of interfacing with the subprogram during execution. If the process to be coded as a subprogram has future utility then it should be coded as an external callable program. If the subprogram is small (contains only a few statements) and if the cost of executing the interface to and from the subprogram would be high, then it should probably be coded as an in-line block of code. In this case a macro should be considered because it saves coding (only the macro reference need be coded). When a subtree is pruned from the design tree in many places because a call is to be made to the subprogram represented by the subtree, each node that was the root node of the subtree becomes the point of call. Each node should be marked as shown in Fig. 3.2.28. The process that is pruned from the design tree should become a tree by itself in the program documentation.

Each segment of code or module in a computer program should be kept small. Program segments that are coded in higher-order languages should not exceed 150–200 executable statements. A good rule of thumb is that the program should fit on one's desk. That restricts the program to about three pages of code. Larger programs may become too complex. There is seldom a case in which a process must exceed the recommended number of statements. Keeping the number of statements small also aids in testing the program because the testing is divided into smaller, simpler problems. There is also a psychological advantage to keeping the program segments small. Every programmer will admit to the satisfaction of completing the development of a program successfully. Suppose that one could develop programs at the average rate of twenty-five lines of tested code per day (the average experienced by FORTRAN programmers is ten). Assume that a program is developed that will consist of 1000 statements. A programmer will need forty days to complete this work. Forty days is eight weeks. If this program had been segmented into subprograms that averaged 125 statements each, the programmer would have had the satisfaction of completing a program once each week instead of waiting for eight weeks. He would also have found that it was much easier to develop the eight small programs instead of the one large program. Experience to date has shown that it takes less time to complete the eight small programs than it does to complete the one large program.

Interfaces

The computer program is only one of the components in a computer-assisted problem solution. The other components are the host computing system (hardware and software), the data structures that contain the data to be processed, and the users of the program. An interface exists between each of these components. The interfaces between the program and the data, the user and the data, and the user and the program were specified during design. These interfaces must now be examined in the environment of the host computing system. In addition, the interfaces between the host computing system and the program, the data, and the user must be developed.

During the design of the functional model of the solution, the data that was to be used as input to each process and the data that would be produced as output from each process was specified. It must now be specified exactly how this data is to be read from input devices, written to output devices, and passed between program segments. As shown in the functional model, the input set of a component contains elements that are output from other components in the model or that are identical to elements of the input set of the parent component. The designer must decide whether the parent process is to read the data and pass it to its subprocesses or the subprocesses will be allowed to read data themselves. The designer must also decide how the subprocesses will pass data between themselves. Data can be written on auxiliary storage devices during one process and read from those devices by other processes. Data may be placed in a global (common) storage area to be accessed during processes, or passed as parameters through a calling sequence, or some combination of these methods may be used. When a large amount of data is to be passed between program segments, it may be necessary to use auxiliary storage. Global storage areas are established for the data that is to be accessed during many processes in a program. The global storage area is defined for the data that is common to a set of processes. The design tree and the accompanying data tree provide the model for creating this global storage area.

Global or common variables should not be used in a computer program without very careful consideration. Once a scheme has been established and the declarations have been placed into the programs, no program of the set that uses this scheme can alter a global or common declaration without affecting every other program. This can make program maintenance very difficult.

The data could have been passed between the program modules or subprograms as parameters in a parameter list. This has the advantage that only the program segments that are modified are affected.

During the adaptation of the program, the designer must also specify exactly how the various blocks of code are to be activated. The program may consist of all in-line code, or a combination of in-line code and external subprograms. The blocks of code correspond to the nodes in the tree and the graph. The designer has already decided which are to be coded in-line and which are to be referenced as subprograms. Now the graph is used to determine the control interfaces. The program segments are either executed in sequence, or they are repeated, or selected from a set of segments. The designer must choose the control statements that are to be used and the procedure for referencing callable external programs.

3.2.14 Summary of Computer Program Design

In Section 3.2 we have discussed the methodology of computer program design. The proposed design method proceeds in stepwise fashion, from the most general to the most detailed level. It cycles through the same type of design tasks for each level of detail. This method produces an "error free" design that can be validated by means of a "proof of correctness." The design is modular and can be documented and reviewed as it is produced, not after the fact.

Our design method considers the computer program, at any level of detail, to be a process. To design this process one has to:

1. Decompose it into subprocesses,
2. Define the output sets of the subprocesses,
3. Define the input sets of the subprocesses,
4. Define the functional correspondence between elements of the input and output set,
5. Define operational sequences of the subprocesses,
6. Define criteria for execution and, within those, dependence of subprocess outputs on inputs,
7. Prove correctness of the design,
8. Verify satisfaction of requirements and evaluate performance,
9. Document, and
10. Review the design.

The design then proceeds from step (1) for each subprocess that was defined at the previous level of detail. This continues until no process or subprocess can be decomposed any further.

We note that the proof of correctness and verification steps are relative to the solution requirements definition and other specified conditions. It is assumed at this stage that the solution requirements definition and other specified conditions are correct. If this is not the case a computer program may be designed properly to carry out an erroneous (e.g., nonconvergent or unstable) algorithm. Such a program design can be proven correct in spite of the fact that the corresponding computer program will not solve the mathematical modeling problem. Thus the proof of correctness is a necessary but not sufficient condition for success of the mathematical modeling project.

The design of all processes and subprocesses in a computer program results in a somewhat abstract model. Before this model can be coded and implemented it must be adapted to the computing environment in which it is expected to operate. This subject is discussed in Subsection 3.2.13, where we show how to make this adaptation.

At this stage of the computer implementation of our mathematical model we have a tested, verified, documented, and reviewed design that is ready for construction and testing as a program. These topics are discussed in Section 3.3.

3.2.15   Example of Computer Program Design

In this subsection we develop the general design of a computer program implementing a mathematical model of steady nonlinear flow in a pipe network. The analytical stages of development of this model are discussed in Section 2.3. Here we follow and illustrate the general top-down design steps that were outlined in Subsections 3.2.1–3.2.12. We pursue the design through the two top levels, and document it following the documentation format outlined in Subsection 3.2.11, first for the first level of design and then for the second level. Data handling subprocesses have been omitted in our illustration so as to focus attention on the algorithm implementation.

1.   First level of design

1. *A mathematical model of steady nonlinear flow in a pipe network.*

2. *Summary of program function and solution method.* This program is a computer implementation of a mathematical model of a steady flow of fluid in a pipe network. The purpose of the network is to transmit fluid from a set of sources (one or more) to a set of consumers (one or more). The purpose of the model is to assist in the pipe network design. The model relates the network structure, element characteristics, flows, and pressures in the network. The modeling problem is defined by a set of nonlinear algebraic and transcendental equations. The solution method is in two parts: one part generates a so-called computational model and the other is an iterative method for solving the set of nonlinear model equations. The mathematical model and the

solution of the modeling problem are general and can be used for many different network flow problems.

3. *Requirement specifications* (see Subsection 3.1.3).

(a) *Basic requirements.* The process to be performed by the computer program consists of the implicit loop procedure and the iterative procedure discussed in detail in Subsections 2.3.4 and 2.3.5 respectively. This discussion need not be repeated here.

(b) *Supplemental requirements.*

(i) Program development cost is to be under $XXXX.

(ii) Program development is to be completed within XX months from initiation.

(iii) Program is to be installed and operational on computer of type XXX, accessible through simple interactive terminals for all its input and output.

(iv) Whenever a single modeling experiment requires 10 minutes for computation, the process should be stopped at the end of the current iteration step. In this case an error message and the current, not final, approximation to the solution should be printed out.

(v) All residuals that are smaller than 0.10 should be neglected.

4. *Output* (see Subsection 3.2.4).

(a) *For each node j, $j = 1, \ldots, m$* a pressure P_j, a floating point number specified in lbs/in² to two decimal places, ranging within ± 1000 lbs/in².

(b) *For each boundary node j in $1, \ldots, m$* a flow $\bar{Q}_j$, a floating point number specified in ft³/sec to two decimal places, ranging within ± 1000 ft³/sec.

(c) For each element $i, i = 1, \ldots, \ell$ a flow q_i, a floating point number specified in ft³/sec to two decimal places, ranging within ± 1000 ft³/sec.

(d) *Error messages:*

(i) Insufficient input data was provided.

(ii) Redundant input data was provided.

(iii) Inconsistencies occur in input data.

(iv) Problem as specified exceeds program storage capability.

(v) Iteration did not converge.

5. *Input* (see Subsection 3.2.5).

(a) *Network structure.*

(i) Number of elements ℓ, an integer less than 1000.

(ii) Number of nodes m, an integer less than 2000.

(iii) For each element $i, i = 1, \ldots, \ell$, two integers—the initial and terminal node of element i, both integers are $\leq m$.

(b) *Element types and characteristic sizes.* For each $i, i = 1, \ldots, \ell$, specify:

(i) Type of element, an integer ranging from 1 to 10, representing element types: pipes, pumps, pressure regulators, reservoirs, etc.

(ii) Element characteristic size, i.e., α_i and a in Eq. (2.3.4), floating point numbers specified to four decimal places ranging between ± 1000. (For example, for each pipe specify length in feet, diameter in inches and pipe flow resistance coefficient such as Hazen-Williams coefficient in Eq. (2.3.4').).)

(c) *Boundary conditions.*

(i) List of boundary nodes, integers $\leq m$.

(ii) For problem of Type I (Subsection 2.3.2): a list of boundary node elevations and pressures, floating point numbers, specified in ft and lbs/in^2 respectively to two decimal places ranging within 0–10,000.

(iii) For problem of Type II (Subsection 2.3.2): a list of boundary node elevations and flows, floating point numbers, specified in ft and ft^3/sec respectively to two decimal places, ranging within 0–10,000 (elevations) and ± 1000 (flows).

(d) *Program user instructions.*

(i) Create data file.

(ii) Load existing data file.

(iii) Modify data file.

(iv) Print input.

(v) Perform modeling experiment.

(vi) Print output.

(e) *Functional correspondence between input and output* (see Subsection 3.2.6). Functional correspondence is specified by equations of mathematical model, Eqs. (2.3.1)–(2.3.5) and boundary conditions. P_j, $\bar{Q}_j$ and q_i are functions of all inputs.

6. *List of subprocesses* (see Subsection 3.2.3). (Note: In this illustration we have concentrated on the algorithm and omitted all data handling subprocesses.)

(a) *Create computational model.* This is accomplished by pursuing the four steps of the "path selection strategy" discussed in Subsection 2.3.4.

(b) *Test residuals* and stop if residuals are below tolerance, otherwise continue to (c) below.

(c) *Solve network flow equations.* This is accomplished by means of the iteration procedure described in Subsection 2.3.5.

(d) *Evaluate residuals.* This is accomplished by using the path sequence established in (a) above and the solution of the network equations established in (c) above.

(e) *Repeat* from (b) unless time > 10 minutes.

7. *Graphical model.* (Note: Only subprocesses listed in Section 6 are included.)

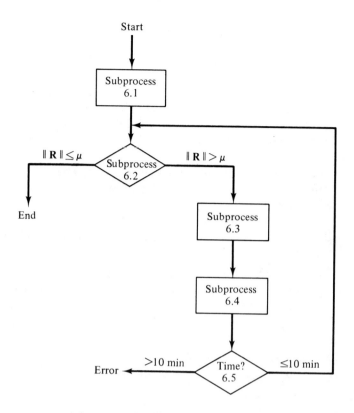

8. *Operational sequences* (see Subsection 3.2.7). There are four operational sequences in the above model:

    (a) $\|\mathbf{R}\| \leq \mu$ (6(b) to end).

    (b) $\|\mathbf{R}\| > \mu$ (6(b) to 6(c)).

    (c) Time $> 10'$ (6(e) to error).

    (d) Time $\leq 10'$ (6(e) to 6(b)).

2.   Second level of design

1. *Names of subprocesses.*

    (a) *Create computational model* (see 6(a), first level).

    (b) *Test residuals* (see 6(b), first level).

    (c) *Solve network flow equations* (see 6(c), first level).

    (d) *Evaluate residuals* (see 6(d), first level).

2. *Summary of functions.* Refer to descriptions in Section 6 of first level.

3. *Requirements specifications.*

    (a) *Basic requirements.* Refer to 3(a), first level.

    (b) *Supplemental requirements.* Refer to 3(b), first level.

4. *Outputs of subprocesses.*

    (a) *Computational model* consisting of five units specified in Subsection 2.3.4:

(i) Element number sequence, a sequence of integers $\leq \ell$.

(ii) Estimated boundary flows, $\bar{Q}_j$ for each boundary node j in $1, \ldots, m$ a floating point number specified in ft³/sec to 2 decimal places, ranging within ± 1000 ft³/sec.

(iii) Estimated element flows, q_i, for each element $i, i = 1, \ldots, \ell$, a floating point number specified in ft³/sec to 2 decimal places, ranging within ± 1000 ft³/sec.

(iv) Pressure residuals, floating point numbers, specified for certain elements i in $1, \ldots, \ell$, in lbs/in² to 2 decimal places, ranging within ± 1000 lbs/in².

(v) Flow residuals, floating point numbers specified for certain nodes j in $1, \ldots, m$ in ft³/sec to 2 decimal places, ranging within ± 1000 ft³/sec.

(b) *Residual test* output is either $\leq$ or $> \mu$.

(c) *Network equation solution* consisting of outputs 4(a), 4(b), and 4(c), first level.

(d) *Residual evaluation* consisting of pressure and flow residuals output specified as in 4(a)(iv) and 4(a)(v) above.

5. *Inputs to subprocesses.*

(a) *Computational model* requires all the input as specified in 5(a), 5(b), and 5(c), first level.

(b) *Residual test* requires the residuals output specified as in 4(a)(iv) and 4(a)(v) above.

(c) *Network flow equation solution* process requires the computational model specified in 4(a) above and all other input specified in 5(a), 5(b), and 5(c), first level.

(d) *Residual evaluation* requires the path sequence specified in 4(a)(i), and the network equation solution specified as output 4(a), 4(b), and 4(c), first level.

(e) *Functional correspondence between input and output.* See 5(e), first level.

6. *List of subprocesses.*

(a) *Create computational model* (see Subsection 2.3.4).

(i) Estimate all but one boundary flow, compute the latter.

(ii) Test whether/not any more input/output calculations can be performed without additional estimates: if yes, proceed; if not proceed from 6(a)(iv).

(iii) Perform all input/output calculations that can be performed.

(iv) Test whether or not every element is in path sequence: if not, proceed; if yes, done.

(v) Find node with known (or previously calculated) pressure and fewest elements that have not been included in path sequence.

(vi) Estimate flow in one of the elements; continue from 6(a)(iii).
(b) *Test residuals.* See 6(b), first level.
(c) *Solve network flow equations.*
 (i) Update Jacobian matrix, Eq. (2.3.14).
 (ii) Solve linear equations (2.3.13).
 (iii) Correct solution estimate, Eq. (2.3.12).
(d) *Evaluate residuals.* See 6(d), first level.
7. *Graphical model.*
 (a) *Create computational model.*

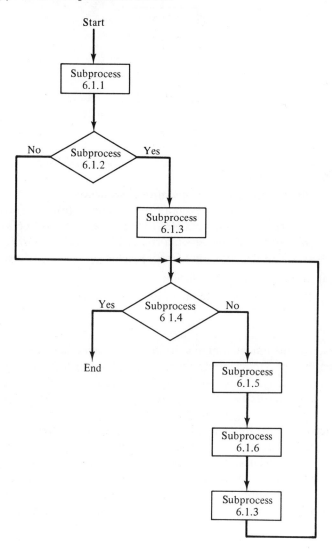

(b) *Residual test* not applicable.

(c) *Solve network flow equations.*

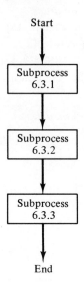

8. *Operational sequences.*

   (a) Process 6(a) has four operational sequences:
     (i) Test in subprocess 6(a)(ii) yields no: (6(a)(i) to 6(a)(iv).
     (ii) Test in subprocess 6(a)(ii) yields yes: (6(a)(i) to 6(a)(iii).
     (iii) Test in subprocess 6(a)(iv) yields no: (6(a)(iii) to 6(a)(v)).
     (iv) Test in subprocess 6(a)(iv) yields yes: (6(a)(iii) to end).

   (b) *Residual test* not applicable.

   (c) Process 6(c) has one operational sequence.

Summary

The above document contains the general design of a computer program implementing the model of steady nonlinear flow in a pipe network. The general design starts with the top level and proceeds to one further level. Data handling subprocesses have been omitted and we have not provided an adaptation in a programming language to a specific computer. This adaptation can be accomplished following each level of design, or after the design has been completed, i.e., developed to the lowest possible level. One further comment is in order relative to computer adaptation of the second-level subprocess 6(c)(ii), solve linear equations (2.3.13); we recommend the use of a subroutine for the subprocess, from one of the subroutine collections listed in Subsection 2.1.5. One may even find a subroutine suitable for the whole process 6(c), solve network flow (nonlinear) equations. In either case, the resulting computer adaptation will be more reliable and effective than an ad hoc program. The proof of correctness of our design is straight-forward; every operational sequence, in both levels, will terminate either by reaching "end" or "error," and if "done" is reached, all required outputs will be produced. Verification of requirements satisfaction will be accomplished following further refinement of the design.

3.3 COMPUTER PROGRAM CONSTRUCTION AND TESTING

Program coding and testing is the last step in the program development process preceding acceptance testing, certification, and subsequent use. The program construction and test plans were formulated in the program design phase. Coding the program should be little more than transcribing the program design in one of the selected programming languages. The final program document and any final changes to the user document, the operator document, the installation document, and the maintenance document are also produced during this phase.

3.3.1 Computer Program Construction

A plan is formulated during the program design phase for the construction of the computer program. By construction, we mean coding the program modules in a programming language. Therefore, this plan defines the minimum configuration of the program, the order of module construction within the minimum configuration, and the order in which the remaining program modules will be added to the minimum configuration. The plan also defines data structures for the program modules.

The minimum configuration of the program is a skeleton, or subset of the program as depicted in the program model. This skeleton program is an executable block of code; however, many of the functions have been deliberately left out in order to simplify coding and testing. A minimum configuration may be a main program with input and output statements, control statements, and references to dummy subprograms. In the case of a completely hierarchical program, where the main program contains only input, output, control statements, and subprogram calls, the main program is exactly the minimum configuration of the program. It can be executed by reading data, selecting various paths depending upon the input data, and writing results to the output devices. In this case the output may be the input data printed for verification, and statements or data printed to indicate which paths are traveled through the program. This simplifies the testing process because it does not involve the functions referenced by the main program.

The dummy subprograms called by the main program can return control to the calling program having done nothing, or they can print messages indicating they have been called. They can also assign specific values to their output parameters based upon the value of their input parameters. Simple table look-up routines may be used in these dummy subprograms. These routines use the input values to select output values based upon the specifications of the subprogram that is temporarily replaced by the dummy subprogram. A very simple table look-up routine should be used; otherwise, the chance of error in the program may be increased. Such dummy routines could be a part of a program test library. The dummy modules should be kept as simple as possible to avoid any chance that they may contain errors. They should be well designed and tested. After the main program has been tested, the actual code for the subprograms can replace the dummy subprograms.

Not more than one dummy subprogram should be replaced with real code at a time. This ensures that only one item in the program changes at any time. If this were not done it could be very difficult to find an error in the program. Suppose that ten program modules were integrated at the same time, and when they were tested they did not work correctly. Where would one begin looking for the program errors? In the case where a single module is integrated into a program at one time, the errors that could occur are much easier to locate and correct.

One of the authors developed an operating system using the techniques that have been described. The following discussion describes the construction of that operating system (see Burner [1973]).

The system developed was a multiple-access interactive computing system on a minicomputer. The system was designed in top-down fashion as discussed above. The top level in the design depicted the user command language, the processes that could be executed, the sequences in which they could be executed, etc. These processes included "logically connecting" the terminal to the system, logging onto the system, issuing commands for the specific processors, editing a file, compiling a FORTRAN program, executing a FORTRAN program, and "signing off" the system. The basic computing system was constructed and tested as the minimum configuration. It contained the terminal reader, the terminal writer, nucleus initialization, the supervisor, routines to place read and write requests from or to the terminal on the read and write queues, the sign-on routine, the command language processor, and the sign-off routine. The editor and the FORTRAN compiler were dummy routines. They could be executed by typing a command EDIT or FORTRAN and the dummy editor or the dummy compiler would be given control. Once given control, these dummy routines would send a message to the terminal indicating that they had received control, but they could do no processing. To exit from either of these programs, the user entered the command QUIT. This scheme allowed the system to be constructed and put into operation in about three months without having to construct and test these two complex programs. Other modules were also left out of the minimum configuration to simplify construction and testing. The potential users of this time-sharing system were able to use it much sooner than if the entire system had been constructed first. During their practice sessions they were learning to use a system while it was being constructed, and they were able to suggest improvements sooner than is usually the case. This reduced the number of changes that had to be made once the system was in operation. Once the minimum configuration was built and tested, the editor and the FORTRAN compiler were constructed. They were also constructed beginning with minimum configurations.

The design tree is an excellent aid for identifying the minimum configuration of the program. The graphs or their equivalents will also be needed. The tree depicts the entire set of subprocesses at all design levels. The minimum configuration is chosen by pruning these subsystems from the design tree. The subtree that is removed is then replaced with a dummy function.

Once the minimum configuration of the program has been selected, the order must be determined in which the parts of that configuration will be constructed and tested. The following principle should be kept in mind when ordering the module construction list: Never code a routine that cannot be immediately used. This will provide the clue to ordering the construction of the program modules. Obviously, the main program must be first. After it has been coded and tested, the next block of code that must be coded, integrated, and tested is one which depends upon no other module for its input. If more than one such block exists, the selection is arbitrary, although that will not often happen. This routine usually produces output that will be used as input by another routine. This construction order can be picked out of the graph for the main program or any program that is being constructed. Clearly, the program is being constructed and tested "top-down" in execution order. This means that a module is not coded and tested with the aid of a "driving" program, then put on the shelf or into a library until it is needed. It also eliminates the possibility of integrating many routines at the same time. Only one new routine is added to the existing correct program. Coding the lowest-level routines first and testing them with the aid of drivers is called "bottom-up" construction. This approach may be used occasionally.

3.3.2 Language Selection and Coding; The Program Format

Coding a properly designed program, or program module, involves little more than transcribing the design into code; but it is a difficult task if the design has not been carried to a sufficient level of detail. Coding is a transcription from one language to another; the design is documented in a mixture of common English and mathematical/logical symbols. The code is in a language that can be translated (compiled) by the computer into a set of (machine language) instructions. These instructions will cause the process of the program to be carried out on the computer.

Prior to coding, a suitable coding language must be selected. Language selection criteria include:

1. Availability of a compiler (translator) and of sufficient documentation and standards at the computer installation at which the program will be coded and used.

2. Applicability of the language to the problem type and availability of features that simplify coding (e.g., debugging aids) and program use for mathematical modeling (e.g., mathematical subprogram libraries).

3. General acceptance of the language by the builder(s) and user(s) of the mathematical model.

Examples of coding languages that are suitable for mathematical models include FORTRAN, PL/1, ALGOL, and time-sharing languages such as BASIC and APL. All these languages are applicable for coding of mathematical modeling problems. The time-sharing languages BASIC and APL are more suitable for smaller and simpler problems. FORTRAN is by far the most widely accepted language. It also has excellent user conveniences such as debugging aids, and mathematical subroutine libraries, and is available on many different large scale and minicomputers. PL/1 is, in some respects, more powerful than FORTRAN and more difficult to learn. It is also not as widely accepted and available. ALGOL and APL are not as widely available as FORTRAN, although both offer some advantages in handling of mathematical problems.

Programs should be produced according to a *standard format*. That makes them easier to produce, read, and maintain. Such a format includes program subdivision into distinct sections, placed in a specific order, and indented to depict program levels. Code blocks or code sections should immediately follow their functional description taken from the program document and be placed in the program as a remark or comment. A program should be considered an extension of the program document.

The sections of a computer program should include (in the order shown):

1. IDENTIFICATION.

2. ENVIRONMENT.

3. SPECIFICATION.

4. PROCEDURE.

COBOL requires a program division similar to this, but FORTRAN, assembly languages, and other programming languages usually have no such requirements.

The IDENTIFICATION section of a computer program should include at least the following items:

1. PROGRAM NAME.
2. AUTHOR.
3. DATE WRITTEN.
4. REVISION NUMBER AND DATE.

The program name should be taken from the name of the program in the document. It can be placed into the program as the title or as a comment. Suppose that a program is being developed to compute the arithmetic mean, standard deviation, and confidence intervals. Assume that the document for that program is titled

A: COMPUTE ARITHMETIC MEAN, STANDARD DEVIATION, AND CONFIDENCE INTERVALS;

In this case the title for the program should be exactly the same. In FORTRAN it would appear as:

```
C    A: COMPUTE ARITHMETIC MEAN, STANDARD DEVIATION, AND CONFIDENCE
C       INTERVALS.
```

Using the same program title in the code as in the documentation establishes a direct correspondence between the document and the program.

The document also contains a description of the purpose, method of solution, and evaluation of the program design. This description should also be included in the program, at least in abbreviated form.

The ENVIRONMENT section of the program should include at least the following items:

1. COMPUTER.
2. OPERATING SYSTEM.

If the program has been developed to operate on more than one computing system, then the environment section should contain comments to that effect. Whenever the program has been developed to take advantage of some special feature of the computing system, the environment section should explain this as well. It will make it much easier for someone who has to implement this program on another computer in the future.

The SPECIFICATION section of the program contains the declarations for variables, the initialization statements, format statement, etc. Some programmers prefer to place the format statements adjacent to input and output statements. The specification section should be well formatted to make it easy to read. It should also be coded in a way that will make it easy to maintain. Crowding too many specifications together in a single statement is not a good idea. In some cases it is advisable to place each specification in a separate statement. The information for the specification part of a program can be found in the program document. Both the program document and the program itself should have specification parts that are similar or identical.

The PROCEDURE section of a computer program contains comments, control structures, and the blocks of code for each subprocess. The procedure section should be kept free of the items in the other three program sections, making it shorter and easier to read.

If the procedure section of a computer program is coded directly from the program design document, it will be an extension of that design document. It will have a single entry point and a single exit. It should also be readable from the top to the bottom without excessive skipping back and forth. It has been suggested that programs constructed using a restricted set of language structures would have these attributes. The removal of the GO TO statement from programming languages has been suggested. This would make it impossible to establish the types of interconnections between components that make programs difficult to code and read. This is not necessary if a program is well structured and uses small modules. Furthermore, it would make coding extremely difficult in the languages that are in use today.

Böhm and Jacopini [1966] proved that an algorithm could be constructed using a minimum set of basic operations. This set included those shown in Fig. 3.3.1. A computer program that consists of these program structures

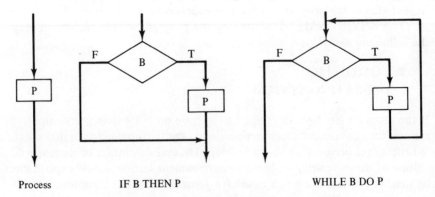

Process   IF B THEN P   WHILE B DO P

Figure 3.3.1 A set of structures sufficient for program production.

reads from the top of the page to the bottom, without having to skip from one page to another. In addition, it is desirable to limit the number of statements in each procedure section. The WHILE statement could cover a large number of pages if the block of code represented by P is long and the IF statement could result in a skip over many pages of code if the number of statements in the block of code represented by P is large. In both cases the size of these program segments should be kept small.

The program will also be easier to read if certain sections of code are indented. Many conventions are recommended for code indentation. We recommend making the code section correspond to the program tree. Indentation is used to show program levels. In Fig. 3.2.3 a program tree is drawn with the root node to the left and the branches extending to the right. The program tree clearly depicts the levels of the blocks of code. To the right of the program tree, structure for the program is shown. Note that the indentations for the code segments are identical to the levels in the program tree.

This method of indentation is also used when a single statement is indented. If it is desired to emphasize the control statements in the program they can be written so that they protrude to the left of the block of code they control. For example,

```
FOR I: = 1 STEP 1 UNTIL N DO
    C(I,J) : = A(I,J) + B(I,J);
```

If the statement under the control of the FOR statement had contained control statements, then they would have obeyed the indentation rules also, but within the statement itself.

The procedure section of the computer program should contain comments explaining what is being done by each block of code. These comments can also be taken directly from the program document. The reader will recall that each program document contains a list of processes. These were written in that document as (refer to Subsection 3.2.11):

```
A.A: READ VARIATES
A.B: COMPUTE ARITHMETIC MEAN
A.C: COMPUTE STANDARD DEVIATION
A.D: COMPUTE CONFIDENCE INTERVALS
A.E: PRINT ARITHMETIC MEAN, STANDARD DEVIATION, AND CONFIDENCE
     INTERVALS
```

These exact statements should appear in the code section of the program as comments. The actual code should immediately follow each comment. For example:

```
C       PROCEDURE SECTION
C       A.A: READ VARIATES
            READ(5,10) N,(X(I),I=1,N)
```

```
C       A.B : COMPUTE ARITHMETIC MEAN
               SUM = 0.0
               DO 100 I = 1,N
               SUM = SUM + X(I)
100            CONTINUE
               MEAN = SUM/N
C       A.C : COMPUTE STANDARD DEVIATION
               SUMSQ = 0.0
               DO 200 I = 1,N
               SUMSQ = SUMSQ + X(I)*X(I)
200            CONTINUE
               STDEV = SQRT(SUMSQ/N - MEAN*MEAN)
C       A.D : COMPUTE CONFIDENCE INTERVALS
               UPPER1 = MEAN + STDEV
               UPPER2 = UPPER1 + STDEV
               LOWER1 = MEAN - STDEV
               LOWER2 = LOWER1 - STDEV
C       A.E : PRINT ARITHMETIC MEAN, STANDARD DEVIATION, AND CONFIDENCE
               INTERVALS
               WRITE(6,20) MEAN, STDEV,LOWER1,UPPER1,LOWER2,UPPER2
```

It is easy to see that if the code were removed from this program, the procedure part of the document would remain. The program is an extension of the document. Using this technique for commenting in the program results in a merger of the program document and the program itself. This technique can be used with any programming language that allows comments, including both assembly language and higher-level languages. An examination of the listings of programs written by good programmers in various languages will show that this type of commenting is common; however, not many of those programmers have striven to make the program an extension of the program document. One can actually lay out the basic program format using the document as a guide. The specification statements are placed in the specification part corresponding to the specifications in the document and the executable statements are placed in the procedure section immediately following the statement telling what the code does. Additional comments may be used to enhance the readability of the program.

The completed program should be reviewed by another programmer. As was the case with the design, it is possible to make errors in coding a program from the design. It is also tempting to "optimize" the code, and this usually results in a deviation from the design. A reviewer can find coding errors and can tell if the code follows the design.

The individual program parts (main program, subroutines, functions, etc.) are coded in the order specified in the program construction plan. As each is completed and reviewed it is integrated into the larger program replacing a dummy block of code. It is tested using the test sets which were specified in the test plan. Coding the program from the design document using standards and methods such as we have discussed, results in a program that is simple to read and is free of design errors. Integrating the program parts according to the program construction plan introduces only a single new code block at a time. This localizes any errors that may exist. Furthermore, any errors that are found should be simple coding errors. The test statements built into the program at the points indicated in the design by the assertions will make it easy to find the errors.

3.3.3 Computer Program Testing

The objective of computer program testing is to demonstrate that the program operates properly or to find errors. Proper operation is obtained if the program is error free. In general, errors can occur in the design, specification, and construction of the program. At this stage we assume that errors of the first two kinds have all been removed and we deal with construction, i.e., structure and coding errors.

A program test plan is also formulated during the design phase. This plan identifies the paths in each program module, the test cases that will cause each path to be traversed, the output expected from each path, and all of the assertions about the execution along each path.

A program cannot be exhaustively tested, yet it must be thoroughly tested to demonstrate that it operates correctly. Each path through the program must be executed to ensure that it terminates and that the expected output is produced. A test set should be derived which will cause each path to be traversed. The examination of the results will tell whether the execution was correct. The program has been very carefully designed to ensure that it is correct. Top-down design has yielded a program in which each module is simple, i.e., contains only a few processes. These processes are simply connected, and only a few paths exist in each module. The design was tested as it was created. The design test used symbolic test cases to yield symbolic test results. Every path in each program module was symbolically executed. Test sets for verifying the correctness of the program code are derived by assigning values to the variables in the symbolic test sets. Each variable is completely defined in the program document. A specific value is assigned to a variable. Associated with each path in a program module is an input vector. Assigning specific values to each variable in that vector will cause that path to be traversed. A special case must be made for each loop in a path. A loop acts like a function. It has a specific input set and it produces a specific output. In addition there is a termination condition for a loop. Each loop in a program consists of initialization of the loop control variable, testing the loop control variable, executing the statements in the body of the loop, and modifying the loop control variable. The loop control variable may be a specific variable independent of the variables in the loop itself or it may be one of the output variables of the loop. Test sets must be selected that will cause the loop termination to be tested and to produce the output of the loop. At least two specific termination conditions should be tested. An input set should be selected that will cause the statements in the body of the loop to be bypassed and another set should be selected that will cause the statements in the body of the loop to be executed at least once. Other loop test sets can be derived to test for specific conditions within the execution of the loop. Each of these test sets can be derived from the symbolic test sets in the program design document for the module containing the loop.

The expected results of the test of each program path can also be derived from the program document symbolic test sets. An output vector is associated with each program path. Actual values have been assigned to the variables in the corresponding input vector for that path. Those values must be used to compute the expected output of the execution of the program path. If an exact computation cannot be performed because of its length, then the range of values for the output variables must be computed from the input values. The expected results must lie within the computed bounds.

Each test of the program must be evaluated. The test plan gives the test input and the expected test results for each path in the program. Test evaluation involves examining the results of the test to ensure that the expected results have been achieved. If the expected test results have not been achieved, then the error must be found. Assuming that the design is correct, the error could exist in the transcription of the design into code, in the derivation of the test cases, or in the derivation of the expected test results. The program design model defines the expected values of every program variable before and after the execution of each block of code. The design asserts that for certain input values assigned to the input vector components along each path output variables, switches, and temporary variables will be assigned specific values or ranges of values. During the testing of a program, statements should be included in the program that print the values assigned to these program variables. This includes printing the values of the variables immediately after the execution of the statements that read input values. The examination of the printout of the values of the variables at specific points in the program will help locate the errors. The input test cases could be incorrect. This will be identified by examining the printout of the input test cases. Examining the values of intermediate variables, switches, and outputs at various points in the program will identify the block of code that is in error.

The program test plan for each module should contain the identification of each path, the input test case for each path, the expected output of the execution of the functions along that path, and the assertions about the state of the variables before and after the execution of the functions along each path.

3.3.4 Summary

In this section we have described the last stage of computer implementation of our mathematical model: computer program construction, coding, and testing. The computer implementation is now ready for use in mathematical modeling exercises.

To be continually useful, computer programs must be maintained and occasionally updated. Updates are program changes and, as such, they must be properly coordinated and controlled. This is accomplished by means of a discipline referred to as "configuration control." Program configuration control and maintenance are the subjects of Section 3.4.

3.4 CONFIGURATION CONTROL AND MAINTENANCE

Once a program module, its documentation, and its test cases have been completed, they are placed under configuration control. This means that any changes to these items must be approved by some one with the proper authority and coordinated with all those who may be impacted by the changes. Programs, documentation, and test cases will be changed during the lifetime of the program to correct errors, to make improvements, and to satisfy changing requirements. Program change is an extension of program development; once a change request has been approved, the actual change is made following the program development procedures discussed in Sections 3.2 and 3.3. Program configuration control and maintenance are the subjects of this section.

3.4.1 Configuration Control

Configuration control is the management process of controlling changes to a computer program module, its documentation, and its test cases. This is an essential process; otherwise, the program and these supporting items can become poorly defined, and thus difficult to use and maintain. There are many examples of programs that have been changed by various programmers until no one knew for certain what the program did, how it was structured, how well it performed, etc. It is eventually impossible to make effective changes to such a program. In some cases programs have had to be discarded and replaced with costly new programs.

A change control procedure should be formal and it should be followed rigorously. We propose here a change control procedure that has been used successfully on many program development projects.

Change control (configuration control) begins during the development of a program. Each program module that has been completed (developed, reviewed, approved) is placed under configuration control. This means that most of the program modules are placed under configuration control before the programming project is completed.

There are usually two or three levels of configuration control. These levels are author control, program management control, and user control. Under "author" we include a designer, a programmer, and a document author. Program management is the management of the computer program development project. We recognize that in many mathematical model implementation projects, all these "levels" may be one and the same person. In this case, some of the coordination effort outlined below is simpler.

The author controls the program and the supporting items during their development. He is responsible for seeing that these items are developed in such a way that they satisfy the program requirements. The author may make necessary changes during the development, or if the author is responsible for one or more other programmers, he may authorize changes to a program module. Changes need to be controlled to completed modules and not modules under development that have not been reviewed and approved. Once a programmer has approved a module, the program may not be changed without his approval. It is very tempting to make a change to a completed program. Every programmer has had a "better idea" after the program has been completed, and most programmers have made unauthorized changes to programs.

When the program module, its document, and its test cases are complete, these items are submitted to project management for review and approval. If approved, they will be placed under configuration control. If not approved, and there are changes to be made, these changes must be made under author-level configuration control. After changes have been made, the program and supporting items are resubmitted to program management for review and approval. If the program and its supporting items are approved, they are placed under program-level configuration control. Once a program has reached this level of control, the author cannot make changes unless the changes are reviewed and authorized by program management. The configuration control procedure used at the program management level is the same as that used by the author of the program.

There are six major steps in a change control procedure:

1. Request a change,

2. Log the request and schedule a review of the request,

3. Review the change request and give authority to proceed or refuse to grant that authority,

4. Develop the change,

5. Review the change,

6. Implement the change in the program.

A request to change a program module, a design, the program requirements, etc., should always be submitted in writing. If possible a standard form should be used. The request can originate with the user, the programmer, or another person who has some interest in the program.

One of the most frequent requests is to change the program requirements. The original program requirements seldom state all of the users' needs. When a program becomes operational, its use often causes the user to think of new requirements or modifications to the original requirements. Change requests may be made to correct an error that has been found in the program. New algorithms may be discovered that could make an existing program more efficient, its results more accurate, etc.

The written change request should always be submitted to the same person or group (change board) that records the request and reviews it. This change request is filed and becomes a part of the development history of the program. The change board considers the requested changes. It usually renders one of three decisions: The request may be denied, the change board may call for estimates of the impact of the change on the program (cost, schedule, etc.), or the request may be approved and authority given to proceed with the development of the change. If the change board has requested detailed estimates, the change will be reconsidered in the context of these estimates. The final decision to proceed with a change must always reside with project management or the user.

If authority is given to develop a change to a program or a document the task will be accomplished using the program development procedures discussed in Sections 3.2 and 3.3. Changes to programs are discussed in Subsection 3.4.2 on program maintenance.

The completed change to a program or a document (or both) is subjected to a review prior to approval by the change board. On the basis of this review, the change board may authorize the change to be implemented in the operational program or request that the development team take certain actions.

The final step in the change procedure is the actual implementation of the change in the operational program. Once this has been completed, the changes to the program, the documentation, and the test cases are distributed to users.

The change control procedure that has been described can be applied to large program development projects or to small programming projects. The steps in the procedure should be followed in either case. The difference is the number of people involved in the procedure.

3.4.2 Program Maintenance

Program maintenance is essentially the development of changes to a computer program and could be called "continued development" because it follows the same procedures that have been recommended for program development. These changes must be authorized, specified, designed, coded and tested, reviewed, and finally implemented in the operational program. Program changes are made to correct errors, to satisfy new requirements, and to improve the program and its documentation.

An additional problem exists for the programmer during program maintenance. The impact of a change to part of the program on the other parts of the program must be assessed. If this is not done there are risks of introducing side effects into the program. There are many examples of changes to programs that have worked, which caused parts of the program to no longer work properly. If the change is to the program requirements the programmer must identify all parts of the program that will be affected by that change. If the change is to improve an algorithm or to correct an error, the programmer must identify all parts of the program affected in these cases. This is another reason for constructing a program with a high degree of modularity. It becomes possible to reduce the side effects of program changes and it is easier to identify the portions of the program that might be affected by a change.

A very useful tool during program maintenance is the requirements tree. This tree should parallel the structure of the requirements document. It identifies each requirement and displays the hierarchical relationship of the requirements to each other. A reference is placed in the tree to the program parts that satisfy each of the requirements. During design, a program design tree is also produced. This tree may or may not be identical to the requirements tree. The program tree identifies all program parts and displays their hierarchical relationship to each other. If the modules in the program tree contain references to the requirements they satisfy, then a two-way reference exists between the requirements and the program. This becomes a very useful tool for assessing the impact of proposed changes to a computer program.

If a change is proposed to an existing program requirement, the requirement number can be found in the requirements tree. The requirement and all related requirements affected by this change can be identified. The corresponding modules can be found in the program tree. Each of these modules must be examined for effects of the change. An entire subtree or some part of that tree may change because of the changed requirement. Suppose the change requires that the result of a computation produce a larger number of significant digits. It may be possible to isolate the program module that produces the result, and change only the design and code for that module. In other cases the change may be spread over many modules. In all cases the affected modules must be identified and the design and code changes made to each.

The addition of a new requirement may fit into the requirements tree as part of a category such as program performance, or it may cause a new branch to be added to the requirements tree. The effect of this change on the program will have to be assessed beginning at the top of the program design tree. The highest-level design will have to be examined to determine whether the change will show at that level. If it does, it shows first as a changed program specification. The change to the program may be made at a lower-level module in the tree. That module must be identified and then the change can be made. The module(s) may be modified or may have to be completely replaced by a new module(s).

When an error is found in a computer program, it is located in one or more modules. The error should be isolated to the highest-level module affected, then its effect as it propagates downward into the program should be assessed. Changes to the program to correct an error must be made at the highest level and propagated as far down into the program as required. As the changes are designed, they must be tested and reviewed level by level as discussed above. The code cannot be written until the design has been tested, reviewed, and documented. After the code has been written and tested, it should be inspected to ensure that it conforms to the design.

An improvement to a program usually involves the incorporation of a better algorithm. The model builder may have found a better method for accessing a table, a better sort algorithm, a method for solving a system of equations that produces better results with less computation time, etc. If a program has been designed with a high degree of modularity an algorithm will be found to be a subtree in the program tree. The effects on other parts of the program may be minimal. If the input specifications of the module do not change, then there will be no effect on the parts of the program that deliver the input to the module. If the results of the computation are greater in significance a computation that follows may have to be changed to accept better input, or an output statement format may have to be changed to print the greater number of significant digits.

After a program module has been modified according to the change control procedure, it must be carefully tested to ensure that it is correct and satisfies its specifications. In addition, the program that contains the changed module must also be tested to ensure that side effects from the changed module have not inadvertently changed the program. A common mistake is to assume that a change to a module will not affect other parts of the program. This has led to many serious problems.

The same procedure that was used during program development should be used to test the module. Both the design and the program must be tested. The program document contains both symbolic and actual test cases. These can be used for retesting. If the specifications for a module did not change, but the algorithm changed, the same test cases can be used. If the specifications changed then new or modified test cases may have to be developed. If the program was changed to correct an error, the test cases were probably also in error; otherwise, the error would have been discovered during testing. In this case, new or modified test cases will have to be developed.

Testing of a program that contains changed modules could involve only the branch of the tree that contains the program. In other cases it may require a complete retesting of the program. This is one of the reasons it is so important to document both the symbolic and actual test cases that were used during program development. It is important that the modified or new test cases also be documented and placed under configuration control.

The changed module and the program that contains that module should also be reevaluated to determine if the requirements are still satisfied. The evaluation should follow the same procedures used during development.

3.4.3 Summary

Operational computer programs will undergo changes for a variety of reasons. These changes must be made using the same methods that have been recommended in this chapter for program development. Changes to programs must be controlled, i.e., they must be authorized and then implemented when it is certain that they meet their requirements and that they have been thoroughly tested within the program. In this section we have discussed procedures for controlling changes to computer programs and have outlined procedures for maintaining programs and documentation. The actual changes to a program and its documentation are made using the same methods discussed earlier in this chapter for creating the programs and their documents.

REFERENCES

BÖHM, C., and G. JACOPINI, "Flow Diagrams, Turing Machines, and Languages With Only Two Formation Rules," *Comm. ACM*, vol. 9, 5, 1966.

BURNER, H. BLAIR, "An Application of Automata Theory To The Multiple Level Top Down Design of Digital Computer Operating Systems," Ph.D. thesis, Washington State University, 1973.

KATZ, S., and Z. MANNA, "Logical Analysis of Programs," *Comm. ACM*, vol. 19, 188, 1976.

KING, J., "Symbolic Execution and Program Testing," *Comm. ACM*, vol. 19, 385, 1976.

KNUTH, D., "Structured Programming With GoTo Statements," *ACM Surveys*, vol. 6, 261, 1976.

RAMAMOORTHY, C. V., "Analysis of Graphs by Connectivity Considerations," *Journal ACM*, vol. 13, 211, 1966.

WARSHALL, S., "A Theorem on Boolean Matrices," *Journal ACM*, vol. 9, 11, 1962.

Suggested Reading:

ALEXANDER, C., *Notes on the Synthesis of Form*, Harvard University Press, 1970.

Chapter 4

Model Validation and Use

At this stage of model building, we have developed a mathematical model and implemented it on a computer. The model has been validated theoretically in the analytical stage of model building as described in Chapter 2. A validated computer implementation of the model has been obtained by means of the process described in Chapter 3. There are still circumstances under which our model may occasionally produce invalid results. The purpose of Chapter 4 is to discuss some of the reasons for this and to inform the model builder, or user, how he might correct or adjust the model to extend the range of validity as much as possible. The purpose of the chapter is also to explain how to test for validity of results obtained during modeling experiments, and how to interpret modeling results in general. By interpretation we mean mapping from model space to prototype space including specifically analysis of modeling errors and of result sensitivity. Figure 4.1 relates Chapter 4 to other parts of the book.

4.1 INTRODUCTION TO MODEL VALIDATION AND USE

Once the mathematical model has been developed, it is necessary to analyze its structure and the data generated by its computer implementation before the model can be considered credible. In this task we distinguish three separate activities:

1. Mathematical verification, which has been accomplished in the analytical stage of model building (Chapter 2),

2. Computer program verification, which has been accomplished during computer implementation (Chapter 3),

3. Model validation.

The purpose of (1) and (2) is to eliminate unintentional logical errors in the model's structure, the mathematical solution algorithm and the corresponding computer program. That is, to make sure that the model is developed and implemented as intended. To some extent this objective is independent of the quality of numerical results or their relation to the behavior of the prototype. Model validation is another stage of modeling; it usually includes:

1. Analysis of the quality of mathematical formulation, solution algorithms and numerical results;

2. Comparison of the numerical responses from the verified model with corresponding responses or measurements recorded from the prototype;

3. A combination of 1 and 2.

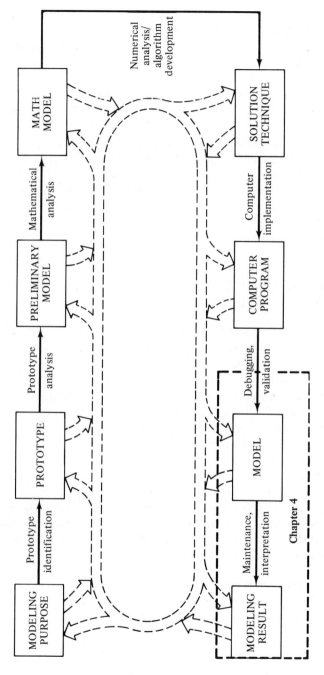

Figure 4.1 Guide to the discussion in Chapter 4.

In Section 4.2 we review some general positions on validation developed throughout the history of science. This section is based primarily on a classical paper by Naylor and Finger [1967]. In Section 4.3 a number of examples is presented which show how unreliable computational results can be even though computer programs implementing the mathematical model are perfectly correct (performing their functions as intended). These unreliable results can be usually attributed to faulty mathematical formulations or poor solution processes.

Other elements of the model validation process are analysis of mathematical error presented in Section 4.4, of other factors contributing to model invalidity discussed in Section 4.5, and sensitivity analysis discussed in Section 4.6. These analyses also assist the model user in the interpretation of his modeling results, that is, in finding the relation of these results to the prototype behavior. Section 4.7 discusses model use and interpretation of modeling results, and Section 4.8 is a conclusion.

4.2 GENERAL CONSIDERATION OF MODEL VALIDITY

Two main positions on validation have developed throughout the history of science and have influenced the methodology of conducting scientific inquiries: rationalism and empiricism. Since both represent extreme views, it is not surprising that practical approaches combine the rationalistic and empirical viewpoints, and seek a reasonable middle-ground compromise.

Rationalists hold that a model is a system of logical deduction from a set of assumed premises which are not open to empirical verification. Consequently validation can be reduced to the problem of identifying a set of fundamental premises which control the behavior of the prototype: If such premises are found and accepted, then the model is valid and no empirical manipulation and comparisons are necessary.

Shannon [1975] gives Forrester's urban model as an example of the rationalist philosophy. Forrester [1969] assumes certain premises underlying his urban models and accepts the quantitative consequences of these principles. Some of his basic assumptions are:

1. If conditions in the urban area are more favorable than those outside, people and industry will move in; of course, the reverse is also true;

2. The changes in housing, population, and industry are the central processes involved in growth and stagnation;

3. The more spent per capita, the greater is the public service offered by the city;

4. The more houses constructed per year in the city, the greater is the city's attractiveness;

5. The larger the percentage of underemployed workers who move up to the labor class per year, the more attractive is the city to outsiders.

Obviously these assumptions can be questioned and changed and a different model may result.

The rationalist view has other serious weaknesses. If we cannot verify the model assumptions by comparing modeling output with empirical data then we have to depend solely on our perception of the basic assumptions made about the prototype and postulated logical relations between them. On the other hand, our knowledge is never complete and absolute. It only represents a series of approximations. Since most scientific laws cannot be stated with absolute certainty, we should look for means of model validation external to the principles of the discipline involved in modeling.

An opposite view is held by empiricists who say that empirical observations are a primary source and the ultimate judge of knowledge. Furthermore, they deny the human brain the ability of recognizing fundamental truths that cannot be independently verified.

Another extreme position worth mentioning is that if the model can accurately predict the behavior of the system under consideration then it is valid. No consideration is given to the validity of the model's internal structure, basic assumptions, or logical relations. This position is clearly unsatisfactory if the purpose of modeling is not only prediction but also understanding of the prototype behavior.

Since none of these positions, if considered separately, can solve the problem of validation, Naylor and Finger [1967] suggested a three-step approach. This three-step validation procedure, although applicable to any model, is particularly suitable for computer simulation models of industrial systems. The three distinct steps are:

1. Determination of face validity of the internal model structure,
2. Empirical validation of the postulated principles,
3. Testing of the model's ability to predict prototype behavior.

Determining Face Validity. In this step we are striving to validate the internal structure of the model using prior research results, existing experimental results, and related theories. We require that the model builder select a set of postulates which can be defended and justified. They are tentative hypotheses and incorporate elements of our best knowledge of the prototype at this time. If the constructed model is substantial in size and complexity, it is conceivable that it will consist of better components (i.e., previously validated) and worse components in which the model builder has less confidence.

Empirical Validation of the Postulated Internal Structure. Here our objective is to verify as many postulated assumptions as possible by subjecting the model components to experimental testing. The theory of statistics concerned with hypotheses testing and estimation is the source of tools for this step.

Ability to Predict the Behavior of the Prototype. This is the final step in which we compare the input-output transformation of the model and the prototype. In some instances, such as modeling hypothetical situations, this type of validation is not applicable. In the majority of physical, industrial, scientific, and socio-economic modeling efforts this step is the ultimate judge of the validity.

There are two predictive validations: historical (retrospective) and forecasting (prospective). Models capable of retrospective predictions are called *replicatively valid.* Those that can in addition forecast, are *predictively valid* models. Their value is higher. Most desirable are the models that are capable of retrospective and prospective predictions, and are *structurally valid* (i.e., understood and correct).

In the comparison of time paths generated by computer simulation models and observed data, a number of statistical techniques can be used. Description or even enumeration of these tests is outside the scope of this book. The reader is referred to the well-established literature on the subject such as Fishman and Kiviat [1968], Fishman [1973], and Van Horn [1971].

4.3 UNRELIABLE COMPUTATIONS

Let us assume that we deal with a complex but mathematically correct model and wish to evaluate results of modeling computations. Furthermore suppose that the computer programs, used to provide the results, have been checked out on an elementary sample problem and have no coding errors, i.e., the programs do precisely what the programmers and the modeler intended. Is it still possible that the obtained results are unreliable or even worthless? The answer is yes. Some of the reasons have been mentioned in Section 2.1.4. Here we illustrate several pitfalls which can considerably diminish the value of output results, or entirely invalidate the modeling effort. As we will see they stem from different sources, but in general, can be attributed to either poor computational procedures or mathematical formulations unsuitable for numerical computing. The examples shown in this section are simple and can be fully analyzed. In many practical problems, the model complexity and the sheer size of computer programs implementing the model and the solution process make it very difficult to find out if something went wrong and that the results may be quite inaccurate. In some instances a thorough error analysis can determine the validity of numerical results and reveal the need for specially designed computational procedures. In others, the model may turn out to be computationally intractable regardless of what solution methods are used.

Models using recursive relations

Many mathematical models are formulated in terms of recursive relationships. The recursive nature of these models' computational solution procedures implies that an error made in one step will be carried over to the next step. This can result in a dangerous error magnification which can contaiminate or invalidate the desired numerical results. Consider the evaluation of

$$I_n = \int_0^1 x^n e^{x-1} \, dx \tag{4.3.1}$$

for $n = 0, 1, 2, \ldots, 15$. Integrating (4.3.1) by parts gives us the recurrence equation relating I_n and I_{n-1},

$$I_n = 1 - nI_{n-1} \tag{4.3.2}$$

Since I_0 is known to be

$$I_0 = 1 - e^{-1} \tag{4.3.3}$$

we can evaluate $I_n, n = 1, 2, \ldots, 15$ by using (4.3.2).

We have used the Hewlett-Packard 25 electronic calculator with 8-decimal digit precision (which is higher than the single word precision of the IBM 360) to perform this calculation. The results are shown in Table 4.3.1. It is seen that at least the last five numbers in the second column are entirely wrong since we should have had

$$1 > I_n > I_{n+1} > 0 \tag{4.3.4}$$

To see what could go wrong, let

$$\epsilon_n = I_n - \hat{I}_n \tag{4.3.5}$$

denote the error in $\hat{I}_n$, that is the difference between the exact integral value I_n and the computed value $\hat{I}_n$. Substituting into (4.3.2) for I_n and I_{n-1} we get

$$\hat{I}_n + \epsilon_n = 1 - n(\hat{I}_{n-1} + \epsilon_{n-1}) \tag{4.3.6}$$

Since our algorithm sets

$$\hat{I}_n = 1 - n\hat{I}_{n-1} \tag{4.3.7}$$

Table 4.3.1 Alternative Evaluation of Eq. (4.3.1)

n	I_n EVALUATED BY (4.3.2)	I_n EVALUATED BY (4.3.10)
0	0.63212055	0.63212055
1	0.36787945	0.36787944
2	0.26424110	0.26424111
3	0.20727670	0 20727664
4	0.17089320	0.17089341
5	0.14553400	0.14553294
6	0.12679600	0.12680235
7	0.11242800	0.11238350
8	0.10057600	0.10093196
9	0.09481600	0.091612292
10	0.05184000	0.083877070
11	0.42976000	0.077352228
12	−4.1571200	0.071773253
13	55.042560	0.066947702
14	−769.59584	0.062732162
15	$11.544937 * 10^3$	0.059017565

it follows that

$$\epsilon_n = -n\epsilon_{n-1}$$

or

$$\epsilon_n = (-1)^n n! \epsilon_0 \qquad (4.3.8)$$

In our case the value of ϵ_0 is very small relative to I_0 and satisfies the inequality

$$|\epsilon_0| \leq 0.5 * 10^{-8} \qquad (4.3.9)$$

but $n!$ grows extremely fast (e.g., $12! = 0.4790016 * 10^9$) and ϵ_n, $n > 10$ dominates the value of I_n.

Suppose now that we rewrite (4.3.2) as

$$I_{n-1} = \frac{1 - I_n}{n} \qquad (4.3.10)$$

and reverse the sequence of solutions assuming that $I_{20} = 0$ and $\epsilon_{20} = I_{20}$.
If we apply (4.3.10) we can see that

$$\epsilon_{19} = \frac{1 - I_{20}}{20} - \frac{1 - \hat{I}_{20}}{20}$$

$$= -\frac{\epsilon_{20}}{20}$$

and in general

$$\epsilon_n = \frac{(-1)^n \epsilon_{20}}{20 * 19 * \ldots * (n+1)} \tag{4.3.11}$$

for

$$n = 19, 18, \ldots, 0.$$

This is much better since the error quickly decreases at each step and we can compute $I_{15}, I_{14}, \ldots, I_0$ very accurately (see Table 4.3.1). It can be shown that reversing the recursion direction is not always helpful. In some problems the catastrophic growth of error takes place regardless of the evaluation order.

The second problem is of a different nature and is taken from the area of optimizing models where we wish to locate a minimum or a maximum of the objective function $f(\mathbf{x})$ subject to some (or no) constraints. A typical step of the iterative procedures used to find such solutions can be described as follows:

1. Given the current approximation $\mathbf{x}_k$, compute the search direction $\mathbf{d}_k$, where k is the iteration number.

2. Find $\mathbf{x}_{k+1} = \mathbf{x}_k + \lambda_k \mathbf{d}_k$ that improves the value of the objective function $f(\mathbf{x})$ and satisfies the given constraints.

3. Update the information required for the computation of the next search direction and repeat from 1).

One of the well-known procedures for unconstrained optimization uses the following specific computations:

1. $\mathbf{d}_k = -H_k \nabla f_k$
where H_k is a matrix and ∇f_k is the gradient of $f(\mathbf{x})$ at the point $\mathbf{x}_k$.

2. $\mathbf{x}_{k+1} = \mathbf{x}_k + \lambda_k \mathbf{d}_k \tag{4.3.12}$

3. $H_{k+1} = H_k - \dfrac{H_k \mathbf{y}_k \mathbf{y}_k^T H_k}{\mathbf{y}_k^T H_k \mathbf{y}_k} + \dfrac{\delta \mathbf{x}_k \, \delta \mathbf{x}_k^T}{\mathbf{d}_k^T \mathbf{y}_k} \tag{4.3.13}$

   where $\mathbf{y}_k = \nabla f_{k+1} - \nabla f_k$

   $\delta \mathbf{x} = \mathbf{x}_{k+1} - \mathbf{x}_k$

It can be seen that the search direction $\mathbf{d}_k$ lies in the range of H_k (it is a linear combination of the columns of H_k) and that the range of H_{k+1} is a subspace of the range of H_k. This implies that $\mathbf{d}_{k+1}$ lies also in the range of H_k. In the optimization process we are interested in locating an optimal solution in the n-dimensional space E^n. If, however, H_k becomes singular, then the set of all subsequent search directions $\mathbf{d}_k$ fails to span E^n and the iterative process will at best generate a suboptimal solution in a subspace of E^n. Here the difficulty can be attributed to implementing the method on a computer with a finite word length. In the idealized case H_k can be computed exactly and if H_1 is nonsingular so are the subsequent matrices, $H_2, H_3, \ldots$.

The singularity of H_k may generate other troubles, such as $y_k^T H_k y_k \approx 0$ which may cause overflow. The consequences of rounding errors made during the computation of $\mathbf{d}_k$ and H_k have been discussed in detail by Murray [1972].

The reader interested in computations with recurrence relations and their numerical properties may like to consult Fox and Mayers [1968].

When do we have a good solution?

This example has been quoted by Forsythe [1970] to illustrate the ambiguity in evaluating numerical solutions of mathematical problems. Consider the mathematical model defined by the set of linear algebraic equations,

$$0.780x + 0.563y - 0.217 = 0 \tag{4.3.14}$$

$$0.913x + 0.659y - 0.254 = 0 \tag{4.3.15}$$

Suppose that we have two different solutions to this system, namely,

$$(x_1, y_1) = (0.999, -1.001)$$

and

$$(x_2, y_2) = (0.341, -0.087)$$

and wish to find out which is "better." One possible way to check is to substitute (x_1, y_1) and (x_2, y_2) into (4.3.14) and (4.3.15), and compute the residuals. These residuals are:

$$(r_1, r_2) = (-0.001343, -0.001572)$$

for the first solution, and

$$(r_1, r_2) = (-0.000001, 0)$$

for the second.

Judging by the residuals only, the second solution is better since it satisfies the linear equations more accurately. If, on the other hand, our criterion of goodness is the closeness to the exact solution, which is $(1, -1)$, then obviously the first solution is better. We may want to apply different criteria to different models. As pointed out by Forsythe, "... the pitfall to be avoided here is the belief that all such criteria are necessarily satisfied, if one of them is."

Ill-condition of certain mathematical problems

Condition of mathematical models refers to the preservation of certain properties of the solution under small perturbations in the model data, or the solution method. We usually define the problem as well-conditioned if small data perturbations cause only commensurately small changes in the solution and ill-conditioned if the changes in the solution are excessive. This property of the model is a function only of the given model formulation, and is quite independent of the method of solution.

A classical example of ill-condition is the linear equations problem of Hilbert,

$$Hx = b \qquad (4.3.16)$$

where the coefficients of H are defined as

$$H_{ij} = \frac{1}{i+j} \qquad (4.3.17)$$

Consider the case where $i, j = 1, 2, \ldots, 4$ and H^{-1} is exactly

$$H^{-1} = \begin{bmatrix} 200 & -1200 & 2100 & -1120 \\ -1200 & 8100 & -15120 & 8400 \\ 2100 & -15120 & 29400 & -16800 \\ -1120 & 8400 & -16800 & 9800 \end{bmatrix}$$

Let us now investigate the impact of some possible perturbations of the right-hand side vector b. We have

$$H(x + \delta x) = b + \delta b \qquad (4.3.18)$$

and

$$\delta x = H^{-1} \delta b \qquad (4.3.19)$$

If the worst combination of signs of $\delta b_1, \delta b_2, \delta b_3, \delta b_4$ happens and these perturbations are $|\delta_k| \leq 10^{-4}$ then δx can be as large as

$$\delta x^T = (0.462, 3.282, 6.342, 3.612) \qquad (4.3.20)$$

This perturbation of the solution may or may not be catastrophic. If for example $b_k = 1$, $k = 1, 2, 3, 4$ and the corresponding solution of $x = H^{-1}b$ is $x^T = (-20, 180, -420, 280)$ then perturbation (4.3.20) is not very harmful. If, however, we solve this system of equations with $b^T = (1, 0.7227, 0.5697, 0.4714)$ and $x^T = (1.162, -0.234, 2.436, -0.560)$ then the perturbed solution $\mathbf{x} + \delta\mathbf{x}$ is totally worthless. We would like to follow here Fox and Mayers [1968] and emphasize "that only the particular problem has a condition" and that the term ill-condition (or excessive sensitivity) should be used in conjunction with the data in the particular mathematical model.

A similar difficulty is exhibited in the problem defined by the differential equation

$$y' = 10y - 10x^2 + 2x - 10 \qquad (4.3.21)$$

with $y(0)$ specified. The solution at $x = 1$ varies with the different starting values $y(0)$, but in this case this variation is excessive, e.g., $y(0) = 1.0$ gives $y(1) = 2.0$, $y(0) = 1.0001$ gives $y(1) = 3.2026$ and $y(0) = 0.9999$ results in $y(1) = -1.2026$. Again a very small perturbation in the problem data, in fact unavoidable in practical applications, induces a solution change of several orders of magnitude greater. If the mathematical model is severely ill-conditioned there may be no point in generating numerical solutions. Rather a mathematical reformulation is required. The reasons are twofold. First, almost every model is based on approximate data. Second, even if this is not the case, in many problems the effect of round-off errors (and these are unavoidable) is equivalent to perturbing the data of the original model.

Ill condition frequently has another adverse effect: It seriously reduces the rate of convergence of certain iterative solution algorithms. This has been found to be the case in matrix algebra and in optimization computations, for example.

Inaccuracy induced by a solution method

Inaccurate results may occur in mathematical models which are fairly well-conditioned, but use numerically unstable solution methods. For example, a system of linear algebraic equations may be well-conditioned but a poor choice of the order of operations could lead to an intermediate set of equations that are ill-conditioned. Consider the equations

$$
\begin{aligned}
-1.732x_1 + x_2 \qquad\qquad &= 0.1 \\
x_1 - 1.732x_2 + x_3 \qquad\qquad &= 0.1 \\
\text{------------------------------------} \\
x_7 - 1.732x_8 + x_9 \quad &= 0.1 \\
x_8 - 1.732x_9 \quad &= 0.1
\end{aligned}
$$

We can eliminate x_1 from the second equation by adding a multiple of the first equation to the second. Repeating this process successively from the top to the bottom of the set we will obtain an equivalent system of equations which is upper triangular, i.e., the equation number i, $i = 1, 2, \ldots, 9$ does not have variables x_j, $j < i$. Such systems can be solved by back substitution, starting from the last equation. If the process of elimination and backsubstitution is carried out using four-decimal-figure floating point arithmetic we obtain the following result (Fox and Mayers [1968]):

$$\mathbf{x}^T = (0.1582, 0.3739, 0.5894, 0.7467, 0.8000, 0.7389, 0.5800,$$

$$0.3656, 0.1533)$$

This solution is obviously very inaccurate since we should expect the symmetry $x_i = x_{9-i+1}$, $i = 1, 2, 3, 4$ (note that the original system of equations is symmetric). There is, however, nothing wrong with this system of equations as a whole, and an accurate result can be obtained by changing the order of the equations in the process of elimination.

Our second example taken from Forsythe [1970], represents a mathematical model defined as a single ordinary differential equation:

$$\frac{dy}{dx} = f(x, y) \tag{4.3.22}$$

We wish to determine $y(x_1), y(x_2), \ldots, y(x_n), \ldots$ given $y(0)$ and $x_{n+1} - x_n = h$. One possible numerical approach to this problem is to determine y_{n+1} from y_{n-1} and y_n by means of the integral

$$y_{n+1} = y_{n-1} + \int_{(n-1)h}^{(n+1)h} f(x, y(x)) \, dx \tag{4.3.23}$$

which can be approximated by Simpson's formula

$$y_{n+1} = y_{n-1} + \frac{h}{3}(y'_{n-1} + 4y'_n + y'_{n+1}) \tag{4.3.24}$$

Since $y'_{n+1} = f(x_{n+1}, y_{n+1})$ the unknown value y_{n+1} is present on both sides of (4.3.24) and a separate computing process is required to satisfy this equation. Let us assume that this can be done and values $y_1, y_2, \ldots$ can be computed using (4.3.24). To illustrate what can happen numerically we consider the equation

$$\frac{dy}{dx} = -y, \quad y(0) = 1 \tag{4.3.25}$$

whose exact solution is $y = e^{-x}$.

Forsythe selected $h = 0.1$ and carried out the computation with $y_0 = 1$, $y_1 = 0.90483742$ using floating point arithmetic with eight decimal places. The numerical results revealed a noticeable oscillation of y_n for $n > 80$ which was quickly growing and resulted in completely wrong values of y_n, such as $y(13.4) = -0.70769248 * 10^{-7}$. The problem is well-conditioned and the integration formulas are accurate. The disastrous results cannot be explained by the accumulation of round-off errors alone. The difficulty can be understood if we analyze the recurrence formula (4.3.24). Since $y' = -y$, we get from (4.3.24)

$$y_{n+1} = y_{n-1} - \frac{h}{3}(y_{n-1} + 4y_n + y_{n+1})$$

or

$$\left(1 + \frac{h}{3}\right)y_{n+1} + \frac{4h}{3}y_n - \left(1 - \frac{h}{3}\right)y_{n-1} = 0 \qquad (4.3.26)$$

A general solution of this equation has the form

$$y_n = A\lambda_1^n + B\lambda_2^n \qquad (4.3.27)$$

where λ_1 and λ_2 can be obtained by solving the quadratic equation

$$\left(1 + \frac{h}{3}\right)\lambda^2 + \frac{4h}{3}\lambda - \left(1 - \frac{h}{3}\right) = 0 \qquad (4.3.28)$$

For small values of h these roots can be approximated by

$$\lambda_1 = 1 - h \qquad (4.3.29)$$

$$\lambda_2 = -\left(1 + \frac{h}{3}\right) \qquad (4.3.30)$$

Substituting these values into (4.3.27) gives

$$y_n = A(1 - h)^n + B(-1)^n\left(1 + \frac{h}{3}\right)^n \qquad (4.3.31)$$

$$= A(1 - h)^{x/h} + B(-1)^n\left(1 + \frac{h}{3}\right)^{x/h} \qquad (4.3.32)$$

since $x = nh$, and finally

$$y_n = Ae^{-x} + B(-1)^n e^{x/3} \qquad (4.3.33)$$

for small h.

The second term is parasitic and $B = 0$ if the initial conditions $y(0) = 1$, $y(0.1) = e^{-0.1}$ are computed exactly, and the iterative process carried out without round-off error. Since these conditions are not satisfied, the second term, although small initially, increased rapidly and overwhelmed the correct solution e^{-x}. The reason for the oscillatory behavior of computed values of y_n can now be easily understood.

In the last two examples inaccuracy has been induced by the solution methods. These solutions can be improved by using carefully designed algorithms.

4.4 ANALYSIS OF MATHEMATICAL ERRORS IN MODELING COMPUTATIONS

In Subsection 2.1.4 we described different types of potential mathematical model errors. In Section 4.3 we have illustrated unreliable computations. In this section we will consider in more detail the three main sources of mathematical errors in the results of modeling computations; these are:

1. Round-off and cancellation errors (effects of finite word length),
2. Convergence errors,
3. Discretization error.

Round-off and cancellation errors

Regardless of the precision used in numerical calculation, round-off and cancellation errors will be made in floating point arithmetic operations. For all but mathematical models using exclusively integer arithmetic, we should expect computational results that are contaminated by round-off errors. These errors can affect the computed results in at least two ways. First there is a possibility that even small round-off errors can be accumulated during lengthy calculations and dominate the final output.

If rounding to the nearest digit is used, the individual errors tend, in general, to cancel out but the standard deviation of the accumulated error increases with the number of operations, leaving the probability of a large final error. If floating point numbers are chopped off, then there is a bias to error in one direction and, in general, the possibility of large final errors is increased.

The second unfortunate phenomenon is the catastrophic cancellation mentioned previously in Subsection 2.1.4. In this case the results may be grossly erroneous even if the computational process is very simple and short. Certain algorithms become numerically unstable because of catastrophic cancellation and produce very inaccurate results. In some cases the cancellation effects can be avoided by rearranging the order of operations.

Higher precision arithmetic can be used to minimize the effects of rounding. The usual computer precision is equivalent of between 7 to 14 decimal digits. It may be increased to a double or multiple precision arithmetic which lengthens the mantissas of floating point numbers. Multiple precision, however, has its price. It requires additional software and increases the required computational time and memory. Single precision arithmetic is sufficient for most scientific and engineering computations, although some special operations, such as vector products, may be computed using double precision. Higher than double precision is used very seldom.

To evaluate the effects of finite word length, analysis of the computational results can sometimes be undertaken. Detailed round-off error analysis techniques have been developed which can be utilized in practical modeling applications. A particularly useful method, called backward error analysis, shows that errors due to finite word length have the same effect as perturbations to the original problem data. If such analysis is feasible we may be able to conclude that the errors caused by rounding are no worse than errors that result from certain data inaccuracies. The problem of errors in the solution is then reduced to the study of the sensitivity of the final solution to perturbation in the model.

To illustrate this approach to error analysis, consider the problem of solving a system of n linear algebraic equations

$$A\mathbf{x} = \mathbf{b} \qquad (4.4.1)$$

Suppose that we first compute triangular factors of A by using a Gaussian elimination process with partial pivoting, and then solve two systems of triangular equations. Let $\bar{\mathbf{x}}$ denote the computed solution of (4.4.1). It can be proven that there is a matrix δA, depending both on A and $\mathbf{b}$, such that

$$(A + \delta A)\bar{\mathbf{x}} = \mathbf{b} \qquad (4.4.2)$$

where $\| \delta A \|_\infty$ is bounded as follows

$$\| \delta A \|_\infty \leq (n^3 + 3n^2)ku \| A \|_\infty \qquad (4.4.3)$$

The value of u is called a machine round-off unit and for most machines it lies between 10^{-6} and 10^{-15}. The value of k does not exceed 8 in practical computations. Relations (4.4.2) and (4.4.3) give no direct estimate of the error in $\bar{\mathbf{x}}$, but are very useful. For example, if all the elements of A already have relative errors larger than $(n^3 + 3n^2)ku$, then it is usually not meaningful to attempt to compute $\bar{\mathbf{x}}$ more accurately.

Convergence error

Many mathematical models require iterative computations which conceptually generate an infinite sequence of approximations to the solution of the problem. The following three issues are usually associated with such iterative processes:

1. Theoretical convergence of the sequence of approximations to the solution,

2. Rate of convergence,
3. Termination of the iterative process.

Only a finite number of iterations can be executed in practice and the computational process is stopped before theoretical convergence can be obtained. The difference between this final approximation and the exact solution is called *convergence error*. Availability of convergence error estimates or bounds would be helpful in deciding when the iterative process should be terminated. For practical problems, however, this information is seldom available. Lacking theoretically justified stopping criteria we usually continue an iterative process until the difference between two successive iterates is less than a preset tolerance. This reasonable criterion can easily fail in the case of slowly converging processes and should be used with caution. After the process is terminated it may be possible to check the error in the final iterate by seeing how well it satisfies the problem data. But as we have seen in the previous section it is perfectly possible that the original problem data are well satisfied by some approximation and at the same time this approximation may be very far, in some sense, from the true solution.

Discretization error

This error arises from the approximation of continuous problems by discrete or finite dimensional problems. Problems that are originally defined as discrete or finite dimensional, such as graph theoretic or combinatorial, are free of this error. As an example we can consider the boundary value problem involving an ordinary second order differential equation.

$$y'' = f(x, y) \quad \text{given } y(a) \text{ and } y(b) \tag{4.4.4}$$

where x is the independent variable, y is the solution to be determined and $f(x, y)$ is a given function of x and y. We assume that this problem has a unique solution.

An attempt can be made to solve this problem by selecting n points $x_1, x_2 \ldots, x_n$ in the interval $[a, b]$ and replacing the second derivatives of (4.4.4) by finite difference approximations. As the result of this discretization we obtain a system of n algebraic equations in unknowns $\bar{y}_1, \bar{y}_2, \ldots, \bar{y}_n$, which are approximations to the true solutions $y_1, y_2, \ldots, y_n$ at the points $x_1, x_2, \ldots, x_n$. The discretization error is a set of differences $\bar{y}_1 - y_1$, $\bar{y}_2 - y_2, \ldots, \bar{y}_n - y_n$ which can be measured in terms of a single number by applying some vector norm.

The analysis of discretization errors is difficult in practice. Ideally we would like to establish that the $\bar{y}_i$ converge to y_i as the number of discrete points x_i tends to infinity. Then we would try to get estimates or bounds for the discretization error for a fixed finite approximation. In practical validation of mathematical models we are likely to encounter serious difficulties in performing such an analysis, although there are a number of helpful theoretical results in various disciplines of applied mathematics.

Further refinement of the discretization (mesh size) is often used to determine whether or not modelin results are dependent on discretization. If the effect of refinement on the results is negligible, the modeler will get nothing in return for the added computation. Unfortunately, it is often impossible to assess the impact of discretization change alone, because errors of the other two types change when discretization changes; more computation may cause increased round-off and convergence errors.

In this section we considered the three main types of errors in mathematical modeling computations: round-off and cancellation, convergence, and discretization. Errors of these types must be taken into account in the interpretation of modeling results. Several ways in which this may be accomplished, or in which model validity may be improved, were pointed out.

4.5 OTHER REASONS
FOR MODEL INVALIDITY

The range of validity of every mathematical model has limitations. These limitations are directly or indirectly implied by the assumptions and logic underlying the model definition.

For example, the mathematical model for hydroelectric power system optimization described in Section 2.4 does not take into account electric transmission losses. They are considered to be negligible in comparison to the losses resulting from nonoptimal operating plans. Clearly, this model is of no use to anyone who is primarily interested in the transmission losses and wants to compare different plans of distributing electrical power to a group of widespread customers. Another limitation in this model results from ignoring the river system dynamics. It is assumed that the time intervals (of $\frac{1}{2}$ or 1 month length) used in the model are sufficiently long for the water released at any storage reservoir to reach the downstream plant within the same time period.

Perhaps one of the most important considerations in constructing valid mathematical models is the selection of appropriate levels of accuracy and sophistication. The well-known MIT model of the world material development (see Meadows et al. [1972]) is a good example of an oversimplified model. This model views the world as one homogeneous system and disregards differences in culture, tradition, economic development and sociopolitical conditions.

With such heavy simplifications, the model has not been able to generate any realistic forecasts or indications for policy makers, although it has succeeded in attracting world-wide attention to some unavoidable near future problems related to technological development. Mesarovic and Pestel [1974] have suggested a world model based on multilevel systems theory. It divides the world into ten different and mutually interdependent regions of political, economic, and environmental coherence. The model is capable of breaking its data down into smaller units at national levels. Furthermore, to make this project even more realistic the model can be exercised in the interactive mode. The input to the computer, (implementing the model) is applied incrementally over time. A user evaluates the effects of previous incremental inputs and selects new inputs. This interaction, between a person who decides on priorities, costs, and risks to be taken and the computer capable of handling large amounts of data and generating a range of options, is a new intriguing facet of the current mathematical modeling technology.

The reader should not assume, however, that the growing accuracy and sophistication of models always increase their validity and usefulness. In fact, very often they lead to limitations upon a model use due to:

1. Increased computational cost,
2. Difficulties in interpreting numerical results,
3. Excessive amounts of input and output data,
4. Substantial effort needed for validating or modifying models.

The following approaches to improve model validity by means of simplification have been successfully used in science and industry (compare with the discussion in Subsection 2.1.2):

1. *Decomposition*, where a model is partitioned into a number of smaller parts which are analyzed and validated separately before the entire model is assembled,

2. *Lumping*, where some model components are aggregated to form a continuous or lumped submodel,

3. *Method of approximating functions*, which involves finding relatively simple functions to approximate complex relationships in the model. Healy and Kowalik 1975] describe such an approach to an airplane preliminary design system. Another practical methodology for condensation of complex, computer based models has been proposed by Meisel and Collins [1973].

4.6 SENSITIVITY ANALYSIS

In the ideal mathematical model all data are exact and solutions, at least in theory, can be obtained with arbitrary precision that is accurate to any desired number of figures. On the other hand, all models representing physical phenomena and exercised on digital computers have either data of limited precision or inaccuracies that are introduced in the solution process, or both. Hence, it is important to know how sensitive our model behavior is to small changes of the model input parameters. Extreme sensitivity could be an indication that the model structure is incorrect or the solution method inadequate.

There is also another reason why sensitivity analysis is usually a part of the modeling process; the experienced modeler seldom confines his interest to a solution which results from a single set of parameter values defining the problem under consideration. He is usually interested in examining the variations in a solution when changes are made in the parameters in model elements.

Mathematical techniques of sensitivity analysis depend upon the discipline of mathematics involved in the modeling process. Here we have selected two types of models, a dynamic system model and an optimization model, to provide a simple illustration of the concept of sensitivity analysis.

A dynamic model

Suppose that our model is defined as a second-order differential equation:

$$f(\ddot{x}, \dot{x}, x, t, a_0) = 0 \qquad (4.6.1)$$

where $\ddot{x}$ and $\dot{x}$ are the second and first derivatives of $x(t)$ with respect to the time variable t, and a_0 is a parameter. If an approximate value of a_0 is established experimentally we may be interested to find how sensitive is the solution of (4.6.1) $x(t, a_0)$ to the perturbation Δa of a_0. Mathematically we would like to investigate the solutions of

$$f(\ddot{x}, \dot{x}, x, t, a_0 + \Delta a) = 0 \qquad (4.6.2)$$

The sensitivity coefficient which measures the influence of the perturbation Δa on x is

$$s(t, a_0) = \frac{dx(t, x_0)}{da_0} \qquad (4.6.3)$$

If the model is steady state, then, the sensitivity coefficient is a function of a_0 only, $s(t, a_0) = s(a_0)$. In unsteady models, s is a function of a_0 and time.

To determine this coefficient we have to solve a sensitivity equation obtained as follows. Equation (4.6.1) is differentiated with respect to a_0,

$$\frac{\partial f}{\partial \ddot{x}} \frac{\partial \ddot{x}}{\partial a_0} + \frac{\partial f}{\partial \dot{x}} \frac{\partial \dot{x}}{\partial a_0} + \frac{\partial t}{\partial x} \frac{\partial x}{\partial a_0} + \frac{\partial f}{\partial a_0} = 0 \qquad (4.6.4)$$

Assuming that $\ddot{x}$ and $\dot{x}$ are continuous functions of t and a_0 we get

$$\frac{\partial \ddot{x}}{\partial a_0} + \frac{\partial}{\partial t^2}\frac{\partial x}{\partial a_0} = \ddot{s} \qquad (4.6.5)$$

$$\frac{\partial \dot{x}}{\partial a_0} + \frac{\partial}{\partial t}\frac{\partial x}{\partial a_0} = \dot{s} \qquad (4.6.6)$$

and Eq. (4.6.4) becomes

$$\frac{\partial f}{\partial \ddot{x}}\ddot{s} + \frac{\partial f}{\partial \dot{x}}\dot{s} + \frac{\partial f}{\partial x}s = -\frac{\partial f}{\partial a_0} \qquad (4.6.7)$$

The solution of this linear equation provides the desired sensitivity coefficient $s(t, a_0)$. If the initial conditions of (4.6.1) and (4.6.2) are identical, the initial conditions of (4.6.7) are $s(0) = \dot{s}(0) = 0$. If, for instance, Eq. (4.6.1) is

$$\ddot{x} + \dot{x} + ax = f(t) \qquad (4.6.8)$$

where

$$x(0) = x_0, \dot{x}(0) = x_1, \text{ and } a = a_0$$

Then (4.6.7) becomes

$$\ddot{s} + \dot{s} + a_0 s = -x \qquad \cdot \ (4.6.9)$$

with the initial conditions

$$s(0) = \dot{s}(0) = 0$$

After solving the full set of equations (4.6.8) and (4.6.9), we can compute $s(t, a_0)\Delta a$ which is a first-order approximation of the difference between the perturbed solution $x(t, a_0 + \Delta a)$ and the unperturbed solution $x(t, a_0)$. The described procedure applies to equations containing any number of parameters.

For further discussion and extensions of the sensitivity concepts in dynamic systems the reader is referred to Bekey and Karplus [1968] and Tomoviĉ [1963].

A static model

In applications of optimization models to problems of economics, engineering, and operations research we are frequently interested in measuring the sensitivity of the objective function to a change in the constraint parameters. To illustrate a typical approach let us consider a linear programming model

$$\text{minimize } f = \mathbf{c}^T\mathbf{x} \qquad (4.6.10a)$$

subject to

$$A\mathbf{x} = \mathbf{b} \qquad (4.6.10b)$$

$$\mathbf{x} \geq 0 \qquad (4.6.10c)$$

where $\mathbf{c}^T = (c_1, c_2, \ldots, c_n)$ is a cost vector, A is an $m \times n, m < n$ matrix, $\mathbf{b}^T = (b_1, b_2, \ldots, b_m)$ an $m \times 1$ vector and $\mathbf{x}^T = (x_1, x_2, \ldots, x_n)$ the sought vector of activity levels. Assume that problem (4.6.10) has been solved and the optimal solution is

$$\mathbf{x}^* = \begin{pmatrix} \mathbf{x}_B \\ 0 \end{pmatrix} \tag{4.6.11}$$

where

$$\mathbf{x}_B = B^{-1}\mathbf{b}$$

and B is the optimal basis consisting of m columns of A.

Now, assuming certain regularity conditions, small perturbations in the vector $\mathbf{b}$ will not cause the optimal basis to change. Thus for $\mathbf{b} + \Delta\mathbf{b}$ the optimal solution is

$$\mathbf{x}^* = \begin{pmatrix} \mathbf{x}_B + \Delta\mathbf{x}_B \\ 0 \end{pmatrix} \tag{4.6.12}$$

where

$$\Delta\mathbf{x}_B = B^{-1}\,\Delta\mathbf{b}$$

The corresponding increment in the cost function is

$$\Delta f = \mathbf{c}_B^T\,\Delta\mathbf{x}_B = \mathbf{c}_B^T B^{-1}\,\Delta\mathbf{b} \tag{4.6.13}$$

where $\mathbf{c}_B$ is an $m \times 1$ vector consisting of elements of $\mathbf{c}$ corresponding to the columns of B. This expression can be further developed if we consider the dual problem corresponding to (4.6.10),

$$\text{maximize } g = \boldsymbol{\lambda}^T\mathbf{b} \tag{4.6.14a}$$

subject to

$$A^T\boldsymbol{\lambda} \leq \mathbf{c} \tag{4.6.14b}$$

whose optimal solution is

$$\boldsymbol{\lambda}^T = \mathbf{c}_B^T B^{-1} \tag{4.6.15}$$

Introducing this solution to (4.6.13) we get

$$\Delta f = \boldsymbol{\lambda}^T\,\Delta\mathbf{b} \tag{4.6.16}$$

Equation (4.6.16) shows that the vector $\boldsymbol{\lambda}$ (simplex multipliers) gives the sensitivity of the optimal function value with respect to small changes in the vector $\mathbf{b}$.

Sensitivity analysis in the broad sense may be applied to changes in constraint conditions as well as to changes in other parameters defining the optimization problem. Its main purposes are:

1. To find how the objective function varies when changes are made in the problem formulation,

2. To study the tradeoffs between gains of the objective function and constraint violations,

3. To evaluate the effect of inaccurate input data on the solution,

4. To detect potential extreme sensitivity which could detract from the model validity.

In general, three types of sensitivity analysis can be applied to mathematical models. First, an analysis whose purpose is to test how sensitive the results or outputs of the model are to the parameter values used in the model.

Second, a sensitivity analysis to find how the results of the model change if the input data vary over a wide range of values. These tests can provide us with information about the scope of validity of the model. Third, and the most involved sensitivity analysis, is aimed at investigating various subcomponents of the model in order to identify those components of the model that are essential and may need further refinement, and the components which have insignificant impact on the model performance and can be simplified or even removed from the model.

As pointed out by Shannon [1975] sensitivity analysis is an important factor in building up users' confidence in a model's validity and is one of the indispensable elements of the model-building process.

4.7 MODEL IMPLEMENTATION AND USE

The last phase of modeling involves model implementation and use of generated results. We assume now that the model has been built and implemented on a computer, and quantitative results are available. It is appropriate at this point to recall that the entire modeling process is iterative in nature (we refer the reader to Fig. 1.4.2 in Section 1.4). It is quite possible that the modeling effort spent so far will be used to gain new insights into the prototype behavior, and an improved or even completely new model will be created. Thus, the range of implementation actions may vary considerably from case to case. One extreme case is where we accept obtained modeling results and use them for practical purposes as planned. Another would be to revise our fundamental objectives and assumptions and create a new model. In practice, the implementation action lies somewhere between these two extremes.

In the implementation stage we should be aware of the fact that every model is only an approximation, and its use depends upon our understanding of the assumptions made in the modeling process. We should carefully review these assumptions, as well as the limits of model validity due to its structure and data, and mathematical simplifications. In addition, there is always the possibility that the model has become obsolete due to new theories or data, or new numerical procedures have become available for improved computations. Resulting observations and objectives can be taken into account in subsequent iterations of modeling, if necessary.

Thus interpretation of the results of modeling experiments is a translation from "model space" to "prototype space." In this process the model user goes, in reverse, through the chain of approximations and simplifications involved in the model and in modeling. He should take into account mathematical errors (round-off and cancellation, convergence, discretization), solution method limitations, data and parameter inaccuracies, and modeling approximations.

It should also be remarked that successful implementation of modeling results may depend heavily upon mutual trust between the modelers and those responsible for using the results of modeling. This is particularly true when models are used to facilitate decision making in such areas as operations research, industrial management, political actions, etc.

Finally, we would like to relate the model uses to the classes of mathematical models and objectives of the modeling effort. In particular, we wish to stress the strong relationship between the model use and the degree of model validity.

Karplus [1977], in one of his papers on the subject of mathematical modeling and system simulation, states, "The ultimate use of the model must be in conformance with the expected validity of the model." He amplifies this point by saying, "In fact, the entire modeling methodology and the eventual utilization of the models is attuned to the uncertainty existing as to the mathematical representation of the system, S, at the outset of the modeling procedure." He also offers an interesting classification of mathematical and simulation models according to the level of uncertainty. Table 4.7.1 shows the large spectrum of mathematical models ranging from white-box type models to black-box type models, with most of the models in the gray area between these extremes.

On the white-box end of the spectrum, we deal with very reliable and accurate models based on well-established physical laws. The white-box models can be constructed almost without experimental verification and guarantee highly accurate representation of the prototypes. The models in this category are used to design products, such as electric circuits, mechanical structures, and the like.

Table 4.7.1 Mathematical Model Uses

MODEL TYPE	EXAMPLE	MODEL USE
White-box	Electric circuit	Product design
	Mechanical structures	Product design
Gray-box	Aircraft control	Performance prediction
	Process control	Performance prediction
	Hydrological	Prediction for action
	Air pollution	Experimentation with control strategies
	Ecological	Test theories
	Physiological	Test theories
	Economic	Gain insight
Black-box	Social	Promote ideas
	Political	Promote ideas

Somewhat less accurate is our model of the hydroelectric system (Section 2.4). We can derive such a model from basic principles of water balance, electric power generation, and geometrical properties of the network. However, some parameters entering the model have to be established experimentally by collecting data related to actual networks. The numerical data obtained are contaminated by unavoidable measurement errors. The usual purpose of such models is to predict the performance of real systems and provide guidance for decision makers. The practical implementation of the model requires substantial experimental verification combined with careful collection of data entering the model.

As we move down the table to models representing ecological, economic, and social phenomena, the representations of the prototypes become less reliable and accurate. Also, the uses of models in this area are entirely different. For example, the large-scale models of national or regional economies are used to gain some understanding of the economic systems, and possibly to develop strategies for practical implementation. Such models are only partly quantitative and their users are quite satisfied if the qualitative behavior of the model is credible.

At the bottom of the table there are black-box type models representing certain aspects of social or political life. In this area, even the fundamental governing principles are not known. Different social and political systems assume quite different relationships between the components and the variables of the systems. The primary purpose of such models may be to promote political ideas and make people aware of certain important social issues; for example, world overpopulation, excessive use of natural resources, or uneven economic development of nations.

4.8 CONCLUSION

To conclude, we point out that in order to validate any kind of mathematical model one has to carefully establish a set of criteria by which to judge model validity. These criteria may significantly differ even among various users of the same model. Thus the validation of a model cannot be separated from the purpose for which it has been designed and from the actual user.

"It may not be true that a model is valid when its potential user says it is valid. It certainly is true that, however valid the model may be in a mathematical or technical sense, nothing will ever come of it unless the user is convinced" (Miller [1975]). This statement is particularly true in areas of application which do not have firm theoretical or experimental bases and where people are asked to make significant decisions on the basis of model predictions.

The reader should also appreciate that the validation of a mathematical or procedural model is always a matter of degree. There are no perfectly valid models. A given model may be more valid by some criteria and le s by others. Well-structured models lend themselves to ite ative refinements which can increase the degree of the model's validity in a desired direction.

However, as the degree of validity of the model increases, so does its development and implementation cost. At the same time the model becomes more complex and this has obvious disadvantages. Thus it may be undesirable to push for the maximum accuracy which is technically possible.

It should also be clear that the process of validation is oft n parallel to the other stages of model building, although a major portion of validation may take place at the end of the process. Each stage of model building involves numerous activities designed to verify and validate certain aspects of the model. At the same time validation is an iterative process, that is, some or all of its activities may be repeated several times during the model-building process.

REFERENCES

BEKEY, G. A., and W. J. KARPLUS, *Hybrid Computation*. New York: Wiley, 1968.

FISHMAN, G. S., *Concepts and Methods in Discrete Event Digital Simulation*, New York: Wiley, 1973.

FISHMAN, G. S., and P. J. KIVIAT, "The Statistics of Discrete-Event Simulation," *Simulation*, vol. 10, 1968.

FORRESTER, J. W., *Urban Dynamics*. Cambridge, Mass.: MIT Press, 1969.

FORSYTHE, G. E., "Pitfalls in Computation, or Why a Math Book Isn't Enough," *Report Number AD 699 897*, Stanford University, Stanford, CA, 1970.

FOX, L., and D. F. MAYERS, *Computing Methods for Scientists and Engineers*, Oxford: Clarendon Press, 1968.

HEALY, M. J., J. S. KOWALIK, and J. W. RAMSAY, "Airplane Engine Selection by Optimization on Surface Fit Approximation," *Journal of Aircraft*, vol. 12, no. 7, 1975.

KARPLUS, W. J., "The Spectrum of Mathematical Modeling and System Simulation," *Mathematics and Computers in Simulation*, vol. XIX, no. 1, 1977.

MEADOWS, D. H., D. L. MEADOWS, J. SANDERS, and W. BEHRENS III, *The Limits to Growth*, New York: Signet, 1972.

MEISEL, W. S., and D. C. COLLINS, "Repro-modeling: An Approach to Efficient Model Utilization and Interpretation," *IEEE Transactions on Systems, Man, and Cybernetics*, 1973.

MESAROVIC, M., and E. PESTEL, *Mankind at the Turning Point*. New York: Signet, 1974.

MILLER, D. R., "Recent Development in Approaches to Validation of Simulation Models," paper presented to the ORSA meeting, Las Vegas, Nev., Nov. 18, 1975.

MURRAY, W., "Failure, the Causes and Cures," in *Numerical Methods for Unconstrained Optimization*, W. Murray (ed.). New York: Academic Press, 1972.

NAYLOR, T. H., and J. M. FINGER, "Verification of Computer Simulation Models," *Management Science*, vol. 14, no. 2, 1967.

SHANNON, R. E., *Systems Simulation—The Art and Science*. Englewood Cliffs, N.J.: Prentice-Hall, 1975.

TOMOVIČ, R., *Sensitivity of Dynamic Systems*. New York: McGraw-Hill, 1963.

VAN HORN, R. L., "Validation of Simulation Results," *Management Science*, vol. 17, no. 5, 1971.

Appendix
Numerical Simulation

A.1 MOTIVATION FOR
NUMERICAL SIMULATION

In Chapters 1 to 4, we have considered mathematical models that can be developed by means of mathematical, analytical, and numerical analysis methods. Frequently, in real life modeling studies, we are confronted with modeling problems which either cannot be cast in mathematical form amenable to analytical or numerical solution techniques, or for which it would be impractical to do so. Despite this difficulty, we may understand or hypothesize how the prototype functions and use a special language to construct a procedural model which, in some sense, behaves like the prototype. This procedural model can be implemented on the computer (analog or digital) and exercised using various sets of data to define modeling problems under consideration. The modeling results can be used to answer questions in prototype space, as in other modeling studies. This process is called *simulation*.

Many of the issues addressed in Chapters 1 to 4 should be of concern to the builder and user of simulation models. The first steps of simulation model development are similar to the first steps of mathematical model development introduced in Section 1.4 and discussed in Section 2.1: prototype identification, statement of the modeling problem, and model definition and analysis. Continuous simulation models (see definition below) involving differential equations should be analyzed to determine their solvability from a mathematical point of view, just like any other mathematical models involving differential equations. Computer implementation, model validation, and use of simulation models depend on circumstances and on the language used. Several of the issues addressed in Chapters 3 and 4 are not relevant to the more strictly procedural simulation models which are constructed by means of the discrete languages mentioned below. The reader would, however, do well to concern himself with mathematical errors generated by simulation models involving successive iterative numerical approximations and similar factors reviewed in Chapter 4. These errors may limit the validity of modeling results or render them invalid, and hence the whole exercise may be useless.

Since the implementation of simulation can be based on computations carried out by analog or digital computers, we distinguish between analog simulation and numerical (digital) simulation. In this Appendix, we are primarily concerned with numerical simulation as opposed to analog simulation where electric circuits are used to study problems defined by differential equations.

Section A.2 introduces two major categories of numerical simulation: continuous and discrete. Section A.3 reviews selected computer languages designed for implementing simulation models on digital computers. This is followed by Section A.4 describing some applications of simulation methodology. (The reader interested in a conceptual and theoretical framework for dealing with simulation models is referred to Zeigler [1976].)

A.2 STRUCTURE OF CONTINUOUS AND DISCRETE SIMULATION MODELS

In simulation we consider the prototype as a collection of interacting elements, the properties of which change with time. We refer to these elements as *entities*, and to their properties as *attributes*. The interactions are called *activities*. We distinguish between continuous and discrete simulation models.

Continuous models consist of differential or algebraic equations in which variables represent the attributes of the entities, and functions represent activities. To solve a continuous model means to find how values of the attributes change as functions of time. Typically, these changes are represented by smooth continuous curves. Continuous models are useful in such areas as engineering design where well-established mathematical procedures give rise to models consisting of differential or algebraic equations. Other areas of application include economics, operations research, and social sciences where we may be interested in long-range trends, or growth and decay behavior. In such studies, many variables can be conveniently aggregated and local discrete fluctuations ignored. Continuous simulation models may also be applied to systems that are discrete in real life but where reasonably accurate solutions can be obtained by averaging values of the model variables. Examples of such approximations are human population studies or mass production by assembly lines. In these cases, we may not be interested in an individual occurrence such as a birth or a finished machine part but wish to know general population trends or the efficiency of a manufacturing process.

Discrete models usually consist of entities and their attributes that change during the simulation runs. These changes occur instantaneously as the simulated time lapses, and are reflected in the output as discontinuous fluctuations. There are many prototypes which can and have to be modeled as discrete, and for which no continuous approximations are valid. An example of such a situation is a model built to study the effects of different computer job scheduling policies. In such a model we may be concerned with queue lengths, waiting times, sequencing of related jobs, and availability of the central processing unit or peripheral equipment, and we cannot ignore the discrete character of the prototype by aggregation or averaging.

The decision as to which type of simulation model to use depends upon the nature of the prototype and the purpose of the simulation study. Tognetti [1969] gives an illustration showing that the distinction between continuous and discrete simulation models is not always obvious. He uses a simple model of population dynamics.

Suppose that we deal with a pure birth process such that the probability that there is an increase by one in a population of n in a small time interval Δt is $n\lambda\,\Delta t + 0(\Delta t)$, where $0(\Delta t)$ is such that

$$\lim_{\Delta t \to 0} \frac{0(\Delta t)}{\Delta t} = 0$$

The probability that there is no change in the population level in the time interval Δt is $1 - (n\lambda\,\Delta t + 0(\Delta t))$, and the probability that there is any other change is $0(\Delta t)$. If the initial population is r then after a time t, the probability of the population having n members, is

$$P_n(t) = \binom{n-1}{n-r} e^{-r\lambda t}(1 - e^{-\lambda t})^{n-r}$$

where $\binom{n-1}{n-r}$ is a binomial coefficient. The mean value of $P_n(t)$ is

$$\mu(t) = re^{\lambda t}$$

The behavior of this expectation can be simulated by a continuous deterministic model

$$\frac{d\mu(t)}{dt} = \lambda\mu(t), \quad \mu(0) = r$$

Consider now the standard deviation of the stochastic process, which is

$$\sigma(t) = \sqrt{r}\, e^{\lambda t}(1 - e^{-\lambda t})^{1/2}$$

and the ratio

$$\lim_{t \to \infty} \frac{\sigma(t)}{\mu(t)} = \frac{1}{\sqrt{r}}$$

If r is very large, random fluctuations about the mean are small compared to the mean value and the deterministic model represented by the differential equation is applicable. If, however, r is small then we have to use the discrete stochastic model. Clearly, for sufficiently small or large values of population size r the choice is simple, but it may be difficult to decide if r is of an intermediate size.

A.2.1 Continuous Simulation Models

In continuous simulation models, the relationships describe the rate at which attributes change as a function of time. Consequently, the continuous model usually consists of ordinary or partial differential equations. The solution of the model describes its evolution with time. Such models have a natural application in engineering problems, in particular those where analog computers have been traditionally used. There are usually two major assumptions made in the process of building and exercising the continuous simulation models:

1. The time scale used in the model is sufficiently large for irregular fluctuations in the variables to be averaged out,

2. The model variables' behavior is smooth enough to be described by mathematical equations.

Problems of this nature can be solved with general-purpose analog computers which consist of various electronic units able to perform mathematical operations. Each problem variable is represented by a computer voltage which can be added and integrated by the direct current amplifiers. These amplifiers (summers, integrators, inverters, etc.) are properly connected to form a circuit such that the computer variables satisfy the same differential equations that control the behavior of the prototype. Analog computers are very fast due to the parallel performance of computations and the fact that the speed of computation is not affected by problem size. To increase their versatility, digital logic and small digital computers have been combined with analog computers giving rise to the so-called hybrid systems (Bekey and Karplus, [1968]). The digital components provide different logical functions and data storage capability. One of the major disadvantages of analog computers is their limited accuracy. Another is the fact that an analog computer is usually dedicated to one problem at a time. The first difficulty may be serious where accurate results are required. Analog computer solutions are measured and recorded and their errors may vary from 0.01 % for digital voltmeters to as much as 10 % for oscilloscopes. These inaccuracies may be unimportant for studies where the model equations are rough approximations of the prototype, or in cases where we are interested primarily in the qualitative nature of the results.

To avoid the shortcomings of analog computers, special languages have been developed and implemented on digital computers. These languages are programs which emulate the behavior of analog or hybrid computers. These computer languages require the user to prepare a block diagram (similar to a diagram needed for the analog computer patchboards) but overcome the problem of inaccuracy of the solutions provided by analog computers. Unfortunately, this gain is achieved by sacrificing speed.

An alternative to a digital computer program (language) emulating analog computers, is an expression-based digital program where a direct description of the model differential equations is provided by the user. Some expression-based languages require that the equations be given in a correct order for calculation; other languages arrange the equations themselves.

To conclude this section, we provide a brief comparison of digital versus analog/hybrid simulation modeling and refer the reader to Ord-Smith and Stephenson [1975] for details.

1. The inferior speed of solution of differential equations is the main disadvantage of numerical simulation as compared with analog simulation. The difference in speed increases with increasing size of the set of equations to be solved. This disadvantage of digital computers may, however, be eliminated by using digital computers with several central processors operating in parallel, or other current fast computer devices.

2. Good solution accuracy is a strong feature of numerical simulation unless it requires excessively short steps in the integrating routines. In such cases we may have to settle for an accuracy comparable with the one given by analog simulators.

3. One of the valuable aspects of analog simulation has been its ability to allow direct man–machine interaction by manual adjustment of the model parameters. Recently, on-line numerical simulation packages have been developed which provide almost equivalent capability.

4. Analog simulation requires scaling of variables in order to keep them within the permitted ranges. The user is relieved from this task in the process of numerical simulation. But, as Ord-Smith and Stephenson [1975] warn, this may be a mixed blessing if the analyzed problem has such potential difficulties as integration instability or function singularity.

5. Numerical simulation has more computational flexibility in expressing complicated nonlinear characteristics of the prototype.

6. In general, it is easier to debug numerical simulation programs than to identify errors and hardware malfunctions in the process of analog simulation.

7. Presently available high-quality computer software, combined with good documentation, favors use of numerical simulation (or hybrid simulation) which can be done conveniently and without the lengthy preparation required by analog simulation.

8. The development and implementation of very large models on analog computers would be impractical because of the unrealistically high cost of such an endeavor.

To sum up, we feel that although there is a certain element of competition between these two methods of simulation, they complement rather than exclude each other.

A.2.2 Discrete Simulation Models

Every discrete simulation model consists of several distinct elements. In order to provide a detailed description of a discrete simulation model, we have collected these in the following four categories.

1. *Components.* A discrete simulation model consists of so-called *entities* representing prototype elements such as machines, people, or messages. These entities may exist in many different states, e.g., they may be busy, idle, or waiting. In the simulation model, the entities interact with one another and an interaction between entities is called an *activity*. These interactions are regarded as *events* in the system which result in changes in the *state of the system.*

2. *Time flow.* Simulation models represent dynamic behavior of a prototype and time is a basic controlling variable. Everything that happens in the discrete simulation model is scheduled according to simulated time. This time can be scaled up or down, and the sequence of simulated events follows the sequence of events which take place in the prototype.

3. *Model images.* During the simulation process, the model is run through a set of states which reflect the prototype behavior. The simulation process can be considered as a time series of model states (images) observed during the simulated time period.

4. *Data collection.* The behavior of the simulated prototypes, the study of which is the main objective in simulation, is usually described by a set of data recorded during the period of simulation. These data reflect the effect of events that have taken place during this period. We may be interested in average values of a state variable, a final system state, or frequency diagrams of state variables, among other things.

We now proceed to describe other aspects of discrete simulation models such as the mechanism of advancing time, alternative modeling techniques, and special data structures.

Time-advancing mechanism

As already mentioned, the controlling variable in simulation is time. Thus, it is necessary to construct, in simulation models, a mechanism for timekeeping. The purpose of timekeeping is twofold: to advance the time of the system, and to synchronize the occurrence of events. One commonly used method of time advancement is to follow the system events which represent changes in system's states. In the interevent periods the states of all model entities remain constant. Since there is no action between events, it is possible to skip over time of inactivity. After all state changes have been made at the time corresponding to a particular event, simulated time is advanced to the time of the next event, and at this time new state changes are made. This technique of time advancement is called the *next-event approach* and is almost universally used by the current computer simulation languages for discrete simulation. An alternative mechanism for timekeeping is referred to as a *fixed time step* mode, where simulation time does not depend on the occurrence of events but is advanced by a fixed increment Δt. In this technique all the events that happened between the simulated times t and $t + \Delta t$ appear to have happened at the end of this interval. If Δt is not properly selected, the results of simulation can be very inaccurate. In order to increase accuracy, a small enough Δt can be selected so that no two events can take place during the interval Δt. Still we cannot distinguish between actual and recorded occurrence times. Also, the cost of a simulation run increases with decreasing Δt. The proper selection of timekeeping mechanism should be based on such considerations as the required degree of approximation, acceptable run time and cost, computer storage available, and access to a computer language.

Shannon [1975] suggests the following criteria to be used as a guide in selecting a suitable timekeeping method. The fixed-increment method is preferred if: (i) events occur in a regular and fairly equally spaced manner; (ii) a large number of events occurs during the simulated time and the mean length of events is short; and (iii) the exact nature of the significant events is not well known.

The next-event timekeeping method is preferred if: (i) the system has long periods of time when no significant events occur; (ii) events occur unevenly and the mean length of intervals between events is long; and (iii) if the accuracy of occurrences is important.

Alternative modeling techniques

The concepts of event, activity, and process give rise to three ways of building discrete-event models. The first is the *event scheduling approach*. Whenever an event is scheduled, a record of this event and its time of occurrence is filed in a special list. When the computer program is requested to find the next event, it searches the list to find and perform the event having the earliest scheduled time. When an individual event takes place, the program executes the steps modifying the system's state, collecting the required data, and proceeds to the next scheduled event. The second approach is the *activity approach*, in which activities or tasks carried out in the prototype are described in terms of the resources they require, how long they take, and how they change the state of the resources they use. In this approach, whenever time is advanced to the next event, all activities are scanned to determine which can be started or terminated. The last method, called *process interaction*, emphasizes the progress of an entity (such as a customer) through a system (such as a bank). In this method, chains of events or *processes* are described together with the ways in which they interact.

As we will see in Section A.3, the existing discrete simulation languages are designed in accordance with these three modeling concepts.

Special data structures

Every simulation language defines a model in terms of entities grouped in classes which have one or more distinguishing characteristics. As time elapses during the simulation run, the number of entities within a class, and the values of parameters defining their specific properties, change. Moreover, certain entities of the same class or different classes may have special mathematical or logical relationships that change during simulation. For example, they may constitute a queue which has to be manipulated (entities are added, dropped, regrouped, etc.) according to a given queuing discipline. In order to retain computational efficiency, a discrete-event simulation language should be able to create, regroup, modify, and destroy different data items without physically moving them in the computer storage. A programming technique that enables such manipulation is list processing, which has been widely used in languages and programs for information storage and retrieval, and management information systems. In this technique, each record consists of a number of contiguous words (or bytes) containing record description, and a pointer identifying the address of the next record in the list. A special word, called a header, placed in a known location, contains a pointer to the first record in the list. Another word, called a trailer, points to the last word of the list. This arrangement allows a program to search the records by following the chain of pointers. If we also arrange a set of pointers starting from the trailer, then a list can be searched in either direction. List processing allows us to perform different data manipulations required of any discrete-event simulation. Typical operations include:

1. Access the ith record of a list and modify or examine its content,
2. Insert a new record between any given two records in the list,
3. Delete a record from the list,
4. Add a record at the end of the list,
5. Combine lists,
6. Divide one list into several,
7. Order records in the list according to the values of their parameters,
8. Determine the record count in a list,
9. Search the list for records with given values of certain parameters.

Using list processing, all these operations can be done efficiently. For example, a record can be added to the list, or dropped from it, by rearranging pointers instead of moving the records. At the same time, there is no need to hold unutilized storage space. Hence, both computing time and space efficiency can be improved.

A.3 COMPUTER IMPLEMENTATION OF SIMULATION MODELS AND THE USE OF SPECIAL-PURPOSE LANGUAGES

In this section we review selected computer languages which are used to implement simulation models on the computer. Virtually every general-purpose computer language can be used for this purpose, but choosing a special-purpose language may greatly increase usefulness of a simulation model and reduce its developmental cost. The general-purpose languages, such as FORTRAN, ALGOL, or PL/1 offer certain advantages, such as considerable flexibility in programming, and availability of compilers on the standard types of computer installations. These languages, however, lack many features that can be conveniently used to specify or manipulate the model or handle nontrivial data structures required for simulation.

Special-purpose simulation languages designed to solve either continuous or discrete simulation problems (some languages can handle both types of problems) are less flexible but offer the user substantial advantages. Some of these are:

1. Ease of unambiguously defining complex simulation models,

2. Availability of special functions and subroutines indispensable or very helpful in simulation studies, such as random number generators, or solvers of differential or algebraic equations,

3. Macro facilities allowing extension of language features,

4. Automatic collection of statistical information pertinent to model evaluation,

5. Capability of organizing the simulated activities in time and space,

6. Ability to handle various data structures, such as linked lists and queues,

7. Dynamic storage allocation,

8. Convenient handling of data outputs and program debugging.

Special-purpose simulation languages are better equipped to deal with dynamic models and for running numerical experiments. On the other hand, the selection of a particular language may bias and influence the structure and the definition of the model being constructed. This mutual interrelationship between a language and a model is typical for many industrial applications of simulation.

As explained in Section A.2, there are two major types of simulation models from the mathematical viewpoint: continuous and discrete. The simulation languages are divided accordingly. Simulation languages developed to assist users in solving ordinary and partial differential equations have been reviewed in an excellent paper by Karplus and Nilsen [1974]. In our brief account of these languages we follow this paper and supplement information provided by Karplus and Nilsen with some recent developments.

Finally, the notion of "nonprocedural languages" has to be mentioned. A standard procedural language, such as FORTRAN, can be used for simulation modeling, but is unsatisfactory in some respects. It requires substantial effort in spelling out the detailed implementation of the solution procedure. In nonprocedural programming, the procedural details are supplied by the compiler, and the problem is defined in terms of the notation of the nonprocedural language. Such nonprocedural languages are limited to specific classes of problems.

A.3.1 Continuous-system Simulation Languages

Continuous-system simulation languages are designed to facilitate modeling and solution of dynamic problems formulated in terms of ordinary or partial differential equations. Within this group we can distinguish languages with different modeling orientation, range of capabilities, and applications. Reference works on the simulation languages mentioned in this section are given at the end of the Appendix. These languages may be classified as follows:

1. Some languages *emulate* the behavior of analog or hybrid computers. These languages enable us to connect and control analog blocks (in the form of subroutines) in essentially the same way as this is done on analog computer patchboards. The digital emulation provides increased precision and automatic scaling of variables. Some of the available emulators are MIDAS, MADBLOCK and CSMP 1130. These languages are useful for solving problems in a manner which emulates the simulations formulated for analog computers.

2. The *expression-based languages* do not follow the block structures of analog patchboards, but handle directly the differential equations representing the model.

(a) MIMIC is a nonprocedural language supplied by CDC and used to simulate the structure and functions required in system simulation. The language has very rigid coding requirements and limited diagnostics, but it inexpensive to run. The solution of differential equations is obtained with the use of a fourth-order Runge-Kutta variable step method. No other user supplied equation solvers can be accepted.

(b) DAREP (Differential Analyzer Replacement—Portable) has been developed at the University of Arizona. Except for a small set of machine-dependent routines, the system is coded entirely in FORTRAN. The program is modular, and uses system-independent FORTRAN-based methods for writing and manipulating solution files. Users can add new integration algorithms or a special library of subroutines. FORTRAN can be used to control the logic of simulation studies and to define special functions. Problem equations are entered in a form resembling mathematical notation, and the associated procedural language is FORTRAN. Input data are provided by format-free data cards. Outputs can be generated on a line printer, or a CalComp plotter can be used. The following integration methods are incorporated: Runge-Kutta Merson, fourth-order Runge-Kutta, third-order Runge-Kutta, two-point Runge-Kutta, Adams two-point predictor, simple Euler one-point predictor, and two methods (Pope's and Gear's) for solving systems of stiff differential equations. The language can be used interactively on PDP-9 computers. The language is available for IBM 360, CDC 6600, and DEC-10 machines.

(c) CSMP/360 and CSMP-III can handle either expressions or block-oriented descriptions. They are preprocessors with FORTRAN as an intermediate language, i.e., they accept a simulation model description and run-time instructions, and produce a FORTRAN program which can be compiled and executed. The user's original program can contain FORTRAN statements and use FORTRAN subroutines. Capabilities of these languages include several integration methods, over 50 simulation functions, and macros, a stiff differential equations algorithm (CSMP-III) and excellent diagnostics and debugging aids. Input/output formats are rich and include various tabular and plot arrangements. CSMP-III can use the IBM 2250 display terminal. According to Karplus and Nilsen [1974], both languages are easy to learn and use but they are computationally expensive. The languages are available for IBM machines.

(d) CSSL-III has been developed in accordance with recommendations made by a Simulation Software Committee formed within the Simulation Councils. The Committee has defined desirable features of a simulation language, related to application areas, programming skills, and digital computers. The language has a powerful capability which can be extended in order to satisfy users' requirements. Like other useful languages, it does have error-control and debugging facilities. Integration can be done with the use of several available numerical methods.

A new improved version of CSSL is called RSSL and has been implemented on the CDC 7600.

(e) PROSE is a general-purpose scientific programming language designed to handle a broad spectrum of problems ranging from numerical analysis to simulation. As a general-purpose language, PROSE provides the full capabilities of procedural languages, including:

    i. Simple arithmetic and algebra,

    ii. Vector and matrix algebra,

    iii. Convenient input and output.

A partial list of capabilities related to numerical computations includes:

    i. Automatic evaluation of first- and second-order analytic derivatives, with no limit on the number of independent or dependent variables or on the complexity of their relationships,

ii. Evaluation of maxima and minima of functions,

iii. Solution of systems of equations, including both algebraic and ordinary differential equations, whether they be implicit or explicit, linear or nonlinear,

iv. Solution of process identification problems, boundary value problems, and optimal control problems,

v. Linear and nonlinear programming subject to arbitrary combinations of linear or nonlinear equality and inequality constraints.

Another uncommon feature of this language is the capability to handle discrete, continuous, or mixed simulation problems. The language runs on Control Data Cybernet.

(f) DYNAMO, a language developed by the industrial dynamics group of MIT, uses first-order difference equations to approximate the modeled continuous process. The dynamic systems are described in terms of levels and rate equations and variables. The level variables characterize the state of the system at any particular time. They accumulate the flows described by the rate equations and variables. These equations and variables define how quickly the levels are changing. Using these two types of equations and variables, it it possible to model many prototypes that can be classified as feedback systems. The DYNAMO compiler accepts the model written in the form of level, rate, and auxiliary equations using special postscript notation. The compiler functions are:

i. Error control,

ii. Sequencing equations according to the structural concepts of system dynamics,

iii. Compilation of the program defining the model,

iv. Execution of the program and generation of output in a tabular or graphical form.

The language has many computational facilities in the form of special statistical and mathematical functions. One of the significant weaknesses of the language is a very primitive and inaccurate integration scheme (Euler's method). In spite of this deficiency, DYNAMO has been extensively used to model and solve numerous information feedback systems. It has great appeal to social scientists because of its conceptual simplicity and ease of use.

The DYNAMO compilers are available for the IBM, B 5500, and UNIVAC 1100 series machines.

(g) ACSL is a nonprocedural language, but procedural FORTRAN can be imbedded in the ACSL program. The translator which is written in FORTRAN, converts the model description into FORTRAN code. A particularly powerful feature of ACSL is its macro facility which permits the user to refer to a section of user codes by a symbolic name. When macro is encountered during the translation process, the code to which it refers is automatically inserted. ACSL provides a library of standard macros which can be augmented by its users.

ACSL runs on the UNIVAC 1100 machines but can be made available on other computers.

(h) EASY is a dynamic analysis language that provides both modular modeling and modular analyses. The EASY precompiler allows complex dynamic models to be easily assembled from predefined modeling components. EASY modeling components are available for common dynamic model elements such as linear transfer functions, saturation, hysteresis, integration, nonlinear function generation, and others. Specialized components are also available for a wide variety of technologies. FORTRAN statements can be included in the model description to supplement the predefined components. The EASY precompiler performs all of the interconnections between model components and provides many tests to help verify the correctness of the model. A schematic diagram of the model is prepared by the EASY precompiler to display the information flow within the model.

Once a system model is constructed with the EASY language, several dynamic analyses can be applied to that model. These include:

    i. Time history calculation (nonlinear simulation),
    ii. Steady state analysis,
    iii. Linear model and eigenvalue calculations,
    iv. Transfer function analysis,
    v. Root locus analysis,
    vi. Stability margin calculations,
    vii. Eigenvalue sensitivity analysis,
    viii. Linear optimal controller design.

Each of these analyses provides efficient proven algorithms for gaining insight into the static and dynamic behavior of the system model. Significant savings in flow and computer time can be realized by this convenient access to a variety of powerful analysis techniques.

The EASY program is written in FORTRAN and runs on CDC machines. Digital Equipment and IBM versions of the EASY program are being developed.

3. There exist several digital *simulation languages for partial differential equations*. Models involving partial differential equations originate in many areas of engineering and physics and can be divided into several distinct classes which require specialized numerical solution techniques. The most common categories of equations are elliptic, parabolic, hyperbolic, and biharmonic equations. The applicable numerical approaches include finite-difference methods, finite-element methods, and the method of characteristics.

Current solution methods for partial differential equations are based mainly on finite-difference methods. In this approach, the time and space variables are discretized and the partial differential equations are replaced by a finite set of difference equations whose number equals the number of grid points. (See Subsection 2.2.4 for illustration.)

These linear algebraic equations can be solved either by numerical iterative methods, such as successive overrelaxation, or by direct numerical methods (Gaussian elimination and equivalent schemes). Due to the recent progress in sparse matrix techniques, the direct solution methods are in many instances more effective tools than iterative methods. They have been used successfully to solve very large sets of sparse linear equations. One of the cited examples is the solution of 30,000 linear equations describing an air pollution model.

The partial differential equation languages currently available are nonprocedural preprocessors whose target language is usually FORTRAN. The user describes the partial differential equation, together with boundary and initial conditions, and the finite-difference grid (see Section 2.2). The linear equations are generated automatically and then solved by an algorithm available in the language subroutine library. The languages differ in scope of applicability and the extent to which the user can affect the choice of solution method.

(a) PDEL is a language capable of solving linear and nonlinear elliptic and parabolic partial differential equations in one, two-, or three-space dimensions, and hyperbolic equations in one-dimensional space. The finite-difference equations are solved by one of the following numerical techniques:

    i. A direct method is used to solve tridiagonal systems of linear equations for problems in one-dimensional space,

    ii. An alternating direction method for problems in higher dimensions,

    iii. A successive overrelaxation method for problems in higher dimensions.

(b) LEANS can solve elliptic, parabolic, or hyperbolic differential equations in one-, two-, or three-dimensional spaces and in orthogonal, cylindrical, or spherical coordinate systems. The user specifies the coefficients of a general equation and in this way tells the system about his equation. The mathematical solution method is based on differential-difference equations. The space variables are discretized and all partial derivatives replaced by finite-difference approximations. The obtained system of ordinary differential equations with respect to the time variable is solved by a user selected algorithm.

(c) DSS differs from LEANS in several respects:

i. It will not solve hyperbolic equations,

ii. All independent variables are approximated by finite differences,

iii. A direct tri-diagonal method is used to solve one-dimensional problems, and the Peaceman-Rachford method is used for two-dimensional elliptic and parabolic partial differential equations.

The language has diagnostic facilities which detect input errors and lack of convergence to a solution.

(d) PDLAN is oriented towards problems in meteorology, and handles systems of simultaneous parabolic partial differential equations. It is a procedural language and the user is required to consider the sequence in which calculations are performed. The user has to specify all finite-difference operators and the mesh size used in the finite-difference method.

(e) FORSIM is a FORTRAN-oriented language, where the user is required to specify the coefficients of a general partial differential equation, as in LEANS. The language is particularly suitable for initial value problems. Partial differential equations are reduced to a system of simultaneous ordinary differential equations and these are solved numerically.

All these attempts to develop software for general classes of partial differential equations indicated a great deal of difficulty in solving wide classes of nontrivial and complex problems. On the other hand, the current state of the art in the area of ordinary differential equations is very advanced. Very capable and robust programs for wide classes of ordinary differential equations exist. It is a tempting idea to further develop computer programs that can serve as an interface and be used to apply the methods for ordinary differential equations to problems formulated initially as partial differential equations after the spatial variables have been discretized. Details of this approach to solving partial differential equations can be found in Sincovec and Madsen [1975].

The reader interested in numerical methods for ordinary differential equations is advised to consult a review by Benyon [1968]. The paper recapitulates the main ideas behind numerical methods for integrating ordinary differential equations and compares various classes of methods. The comparison includes considerations of numerical stability, accuracy, and computing time. An extensive list of references is given.

In many models of physical systems, the resulting differential equations are stiff which may cause numerical troubles. Techniques and programs to integrate such systems are described in Gear [1971], Hindmarsh [1972] and [1973]. A reliable and efficient method for dynamically changing integration step size and method order to maintain numerical stability and user specified accuracy can be found in Krogh [1973].

A.3.2 Discrete-system Simulation Languages

Discrete-system simulation languages can be conveniently divided into four main categories:

1. Event oriented,
2. Activity oriented,
3. Process oriented,
4. Transaction-flow oriented (this is a subcategory of (3)).

In the *event oriented* language, each event is represented by an instantaneous occurrence in simulated time and must be scheduled to occur when a proper set of conditions exists.

The languages in this category are used to model processes that are characterized by a large number of entities. The two outstanding representatives of this group of languages are:

SIMSCRIPT II, which runs on IBM 7090, 7094, 360, 370; Honeywell 600/6000; UNIVAC 1107, 1108; GE 625, 635; CDC 3600, 6000, 7000; RCA 70, 45, 55, and

GASP, a FORTRAN IV based language that runs on machines with a FORTRAN compiler.

In the *activity oriented* languages, the discrete occurrences are not scheduled in advance. They are created by a program which contains descriptions of conditions under which any activity can take place. These conditions are scanned before each simulation time advance and if all necessary conditions are met, the proper actions are taken. According to this concept, the activity oriented languages have two major components: a test component and an action component. Shannon [1975] suggests that an activity oriented language should be considered for use if he model has the following characteristics:

1. The model uses a next-event or variable time increment type of timing,

2. The process simulated is highly interactive but involves a fixed number of entities with events happening irregularly,

3. Event occurrence is controlled by cyclic scanning activity programs.

Languages available in this category are:

CSL, which runs on IBM and Honeywell machines,
MILITRAN, which runs on IBM machines.

Process oriented languages deal with models defined as series of events (processes) and require programming of the explicit interaction between the processes. The languages that represent this approach are:

SIMULA 67 runs on UNIVAC 1107, 1108; Burroughs B 5500; CDC 3000 and 6000 series; IBM 360/370.

ASPOL runs on CDC 6000 series. ASPOL has its roots in SIMULA and has been developed by CDC for simulation of computer operating systems.

SOL runs on UNIVAC 1107, 1108.

Transaction-flow oriented languages use the concepts of processes but the flow of activities (called transactions) pass through specially defined blocks. The prototype is represented by a flowchart consisting of language blocks. The program creates transactions, executes them in the blocks and moves them along the flowchart. Writing a program is reduced to specifying a flowchart representation. The transaction-flow oriented languages are easy to use at the expense of a loss of flexibility. A language representing this category is GPSS, which runs on IBM, UNIVAC, and CDC machines.

Three of the simulation languages mentioned in this section have acquired wide acceptance and popularity among scientific and industrial users. These three languages are SIMSCRIPT (event oriented), SIMULA (process oriented), and GPSS (transaction-flow oriented). We give now an overview of these languages which should help the reader to see the differences between their capabilities and ways of expressing simulation models.

SIMSCRIPT II.5

SIMSCRIPT II.5 can be viewed as a general programming language having extra features for digital simulation. It can be learned and used at different levels of sophistication. Level 1 of SIMSCRIPT II.5 is comparable to a very simple algorithmic language such as BASIC. The inexperienced user would appreciate such features as free-form data and programming, simplified output options, and automatic mode conversion.

The next level can be described as a FORTRAN-plus language. Additional capabilities can be found primarily in the area of data structures. The language allows construction of piecemeal arrays, such as two-dimensional arrays with a different number of elements in each row, and complex tree structures. Every program consists of a main program and subroutines. The latter are recursive. Input-output capabilities include those of FORTRAN, but an extensive report generator is available.

Level 3 of SIMSCRIPT II.5 provides ALGOL-like (or PL/1-like) statements. The first three levels view SIMSCRIPT II.5 as an algebraic procedural language.

Level 4 gives SIMSCRIPT II.5 the ability to define and manipulate entities, attributes, and sets. An entity is a program element similar to a subscripted variable. When an entity is created, it can be interpreted as a generic definition for a class of entities. Individual entities have values called attributes which define a particular state of the entity. Attributes are named, not numbered. For example, we may define EMPLOYEE to be an entity, and AGE and SALARY as attributes of EMPLOYEE. Each EMPLOYEE has his own AGE and SALARY attributes. Entities can be of two types: permanent (specified for total duration of program execution), and temporary (specified dynamically as the program proceeds). Entities can be collected together into sets. Commands are available to manipulate these sets; for instance, create an entity, file in a set, remove an entity from a set, search a set, etc. There are also special commands to establish different set disciplines, such as first-in, first-out, or last-in, first-out, or ranking set elements that depend on some attributive values.

The primary task of level 5 is keeping track of simulated time and organizing subprograms that represent system activities. These activities are the main components of a simulated system, and have two properties: they take time, and potentially can change the system's state. The basic concept used to accomplish these tasks is an *event*. An event is an instant of time which marks the beginning or the end of an activity. After an activity is put into operation, the SIMSCRIPT II.5 program schedules an ending event which will terminate the activity. A special training routine executes events at specified future times and keeps track of their order of occurrence. Events can be generated and scheduled internally or externally (triggered by event data cards). The scheduling of events is usually accomplished by using statistical sampling subroutines. SIMSCRIPT II.5 provides eleven functions for generating pseudorandom samples from commonly used statistical distributions. The language also provides routines for collecting statistical information from a simulation run. A FORTRAN subroutine can be invoked from a SIMSCRIPT II.5 program. The language has some capability to detect programming errors by the compiler and the runtime system.

Finally, we mention two extensions of the SIMSCRIPT language: QUICKSIM and ECSS. QUICKSIM is a block-oriented language written in SIMSCRIPT. Over twenty defined blocks are available in this language and additional ones can be written in FORTRAN or SIMSCRIPT. ECSS is a preprocessor with SIMSCRIPT as its target language. The purpose of this language is to simulate computer systems.

The growing list of SIMSCRIPT uses includes models related to urban growth, hydroelectric planning, transportation systems, election redistricting, cancer and tuberculosis studies, hospital planning, communications, and multicomputer networks.

SIMULA 67

SIMULA 67 is a superset of the general algorithmic language ALGOL 60, a language that has an extensive user basis particularly in Europe. Since the arithmetic facilities are the same as in ALGOL 60, SIMULA 67 can be used as a general-purpose programming language. The basic simulation concept in SIMULA is the process, a combination of a data structure and an operation rule. The operation rule specifies the actions to be taken, and the data structure defines the properties of the process. The operation rule consists of events taking place at one instant of simulated time. There is also a mechanism that activates and deactivates events in the process. In a simulation study we may deal with several processes that have the same data structure and operation rule but differ in the values of attributes related to the data structure. Such a group of similar processes is called the process class.

In the SIMULA program, the definition of a process class is accomplished in a block, which is SIMULA's fundamental way for program decomposition. A block is an independent, self-contained part of the program. The data structure is defined as a part of the block head, and the statements of the block define the operation rule. We may look at a block as being a formal description of an aggregated data structure and of the actions defined by the block statements. When a block is executed, a block instance is generated. SIMULA is capable of generating several instances of a given block which may interact or exist at the same time. The need to manipulate block instances and relate block instances to each other makes it necessary to give names to the individual block instances and structure them. SIMULA does it by an extensive list-processing capability.

In the broad repertoire of SIMULA statements there are special simulation oriented statements that can postpone actions, render the process inactive, etc. SIMULA's program is a set of blocks that organize the total execution as a sequence of active phases related to different block instances.

Like other simulation languages, SIMULA has random number generation facilities and algorithms collecting information about simulation runs. Output facilities exist that allow editing of output reports. Good introductory booklets on SIMULA have been written by Hills [1972] and [1973].

SIMULA is a general-purpose language that has been applied to a great variety of scientific and administrative problems. The following list is a small sample:

1. Teaching structured programming,
2. Road traffic behavior,
3. Computer network design,

4. Simulation of computer systems,

5. Military conflicts,

6. Geometric data description language including a preprocessor written in SIMULA,

7. Input to linear programming packages,

8. Scheduling of maintenance work in complex plants,

9. Movement of trains over the railway network in a steel plant,

10. Coordination of the production facilities of the various manufacturers to satisfy market requirements while minimizing costs.

Prototypes analyzed with SIMULA include: factories, warehouses, harbors, social systems, biological systems, population models, telephone exchanges, marketing systems, and control systems. SIMULA has also been used for file-handling problems, text processing, syntax and semantics processing, implementation of language processors, etc.

GPSS V

GPSS simulates prototypes represented by a series of interconnected blocks drawn from a set of specific block types. Each block has a name and is designed to perform actions required by the prototype. There are 48 block types to choose from, each of which can be used repeatedly.

Moving through the system of blocks are entities called in GPSS transactions. Examples of transactions are customers, messages, machine parts, vehicles, etc. Typical block types are:

1. GENERATE, creates transactions,

2. QUEUE, creates a queue of transactions and maintains certain queueing statistics,

3. TABULATE, tabulates the time it took the transaction to reach that point from the time it entered the simulated system,

4. TERMINATE, destroys transactions and eliminates them from the system.

The sequence of events taking place in the prototype is simulated by the movement of transactions from block to block. There can be many different transactions in the block diagram but at any given instant of time each of them is located at some block. A block can contain and act upon one or many transactions simultaneously. Usually transactions are temporary entities that are generated, moved between blocks, and removed by the GPSS program. Some transactions are permanent and remain present in the simulation system throughout the entire period of the simulation run. Blocks that can handle only one item at a time are called *facilities*. An example of a facility is a machine or person. A related transaction may be a job that has to be performed by the machine or person. A *storage* represents another block type that can be used by one or more transactions simultaneously. Examples of storages are groups of machines or a warehouse.

GPSS handles the advancement of time by a block called ADVANCE. When a transaction enters this block, an action time is computed and added to the current time to produce a block departure time. When the time reaches the departure time, the transaction will be moved, if possible, to the next block in the chart. Transactions might possess certain attributes which can be used to make logical decisions within a block. The program has also the ability to collect statistical data about the simulated process, e.g., the current number of transactions in a storage or the length of a queue.

Simple mathematical calculations can be carried out using GPSS's variable statements. There are no elementary mathematical functions such as the trigonometric or logarithm functions. GPSS can generate several basic random number distributions.

Program output can be produced in a standard form without any specific request from the user, or it is possible to create nonstandard output reports according to user specifications. Written tabular reports, as well as graphs, can be generated.

Programming errors can be detected by the assembler or during simulation execution. Minor potential errors are communicated to the user by the system generated warning messages.

A very fine book on GPSS V (the recent version of the language) has been written by Gordon [1975].

As our reader has probably observed, the GPSS language is suitable when studying inventory and queuing type problems. The language is easy to learn and use even for a large-scale system involving these two problem categories. The language cannot handle any sophisticated or accurate computational problems of a mathematical nature nor can it be extended to suit specialized purposes.

A.3.3 Evaluation and Comparison of Discrete Simulation Languages

As we have seen in the preceding section, a number of discrete simulation languages exist that represent different approaches to event scheduling. They are based on rather distinct key concepts and offer various facilities to aid in the process of formulating, programming, and running the simulation model. Clearly, the proper choice of a simulation language is heavily problem dependent. In spite of this model-language interrelationship, it is possible to suggest some general criteria that should be satisfied by a good discrete simulation language. These criteria may also serve as a basis of comparing the existing languages. We will first list criteria for evaluating simulation languages and then use them to compare the three leading languages: SIMSCRIPT, SIMULA, and GPSS. A set of criteria for simulation language evaluation has been established by several authors, among them Kleine [1971], Emshoff and Sisson [1970], Palme [1970], Rogeberg [1973], and Virjo [1972]. A good account of users' views of simulation languages can be found in Shannon [1975].

For the purposes of this discussion, we describe the main criteria that have been suggested. The order in which they are listed is not necessarily significant. The criteria can be grouped under four headings (Virjo [1972]):

1. *Language generality and programming power.* The language should be able to handle a wide area of applications. This implies that it has to offer strong algorithmic capabilities, as well as special simulation oriented features. Among them are numerical computing, simulation concepts for system description, capability to handle sets and queues, random number and random variates generation, performance data collection and flexible data displaying. We have pointed out the language generality, but see also good reasons for developing and using very specialized and limited simulation languages as long as the user can fit his problem into the scope of the language he intends to use. Such limited languages may improve modeling efficiency.

2. *Ease of use.* Here we mean: natural translation of the modeling problem to a corresponding computer program (model closeness), reduction of debugging effort, convenient documentation, and flexible design of experiments. One of the benefits of model closeness is a high degree of readability, ease of modification, and reduced need for documentation. The reduction of the debugging effort is related to language definition as well as quality of implementation. In general, we would like to see that most of the programming errors are detected in the compilation phase and those discovered by the run-time system trigger meaningful diagnostic messages. Good debugging facilities are very important in the simulation language implementation, mainly because of the extensive use of dynamic data structures.

The design of experiments should be flexible in the sense of combining many experiments into one computer run, and convenient reporting of numerical results. Last, but not least, is the requirement of language simplicity. Here we can only require a good balance between the language conceptual richness and the relative ease of learning and using these concepts.

3. *Machine efficiency.* Usually the cost of simulation model development and coding is a major part of the cost of simulation studies. Nevertheless, machine efficiency is important in those cases where many multiple runs are required or the simulation program is used over an extended period of time. Machine efficiency is to a great extent a property of the quality of compiler

implementation which, in turn, depends on the language structure and content. Two common indicators of machine efficiency are execution time and memory requirement.

4. *Availability*. This is a very practical aspect of any computer language. The language must be available on the most common type of computers in order to be useful. It is also very desirable that the language is machine-independent, i.e., the compilers implemented for different machines should be compatible so that the source programs can be run without significant modifications.

It is difficult to provide an unbiased comparison of available languages. One reason is our unequal experience with different implementations, another is subjective documentation provided by the language or computer manufacturers. For these reasons, we have decided to quote some conclusions and numerical results of a comparative study conducted by Antti Virjo [1972] of the Finnish State Computer Centre. Virjo does not represent any language producer and has tested different versions of GPSS, SIMSCRIPT, and SIMULA using three simulation problems: a simple queuing model of an N-machine workshop, an operation of a telephone system (see Gordon [1969]) and a stochastic PERT network (Program Evaluation and Review Technique). The last problem is the most complex and requires good algorithmic facilities (see Pritsker and Kiviat [1969]). For each test problem, several runs were made and average values have been reported. The following statistics have been recorded: central processor time, amount of fast memory used, and the number of input-output requests or disc operations. Virjo ran his experiments on an IBM 360/50 under operating system OS, and a UNIVAC 1108 under EXEC 8. He compared the following versions of simulation languages: SIMSCRIPT I.5, SIMSCRIPT II, SIMULA 67, SIMULA I, GPSS 360, GPSS II, and GPSS 1100. Conclusions based on his results can be summarized as follows:

1. SIMSCRIPT I.5 programs run faster than GPSS 360 Version 2 on IBM 360/50, but use more core memory,

2. SIMSCRIPT II programs run faster for the PERT network than SIMSCRIPT I.5 programs (IBM 360/50),

3. On the UNIVAC side, SIMULA 67 has been consistently the best performer. The second place has been shared between SIMSCRIPT I.5 and SIMULA I. The GPSS programs have been considerably slower.

Virjo [1972] concluded that, "It is difficult to place the three languages in a definite order of preference. The opinion of the author is, however, that as a general purpose language, SIMULA 67 is in many respects superior to the other possible choices."

He remarked, however, that GPSS and SIMSCRIPT have much broader user bases due to their strong marketing organizations. It may also be added that one of the significant obstacles in propagation of SIMULA is the rather expensive compiler rental or purchase. Virjo has also offered some general judgments, based on his experience and other simulation studies, related to the languages he compared.

GPSS

1. Restricted to simple queuing problems,
2. No language extension possible,
3. Easy to learn and use, provides simple conceptual framework,
4. Good debugging facilities, including comprehensive diagnostics,
5. Inflexible input-output, output statistics always generated,
6. Poor computational facilities,
7. Execution efficiency is often poor since GPSS is an interpretative system.

SIMSCRIPT

1. The most machine-independent general-purpose simulation language,
2. Algorithmic capabilities comparable with ALGOL or PL/1, including built-in functions and distributions,
3. Extensive general-purpose programming facilities,
4. Simulation concepts are relatively few and very general,
5. Data collection facilities are excellent,
6. Input-output facilities are good (similar to those of FORTRAN),
7. Poor debugging facilities, noncomprehensive diagnostics, debugging often requires core dump,
8. Security (error detection) is low, compiler allows confusing execution errors,
9. Flexibility of experimental design is quite good,
10. Execution efficiency is high.

SIMULA

1. Algorithmic capabilities comparable to ALGOL;
2. Language extension possibilities allow the construction of special problem-oriented languages,
3. SIMULA is a complex language, but once mastered is easy to use,

4. Very good debugging features,

5. Collection of statistical data on system behavior quite adequate but inferior to SIMSCRIPT II,

6. Input-output facilities comparable to SIMSCRIPT II,

7. Good program readability and structuring,

8. Possibilities for experimental design comparable to SIMSCRIPT II,

9. Machine efficiency of SIMULA 67 compiler (for UNIVAC 1108) is at least as good as the efficiency of the corresponding SIMSCRIPT I.5 compiler.

Another study has been reported by Tognetti and Brett [1972]. They have compared SIMSCRIPT II (IBM 360/50) and SIMULA 67 (CDC 3300) and concluded that although SIMSCRIPT II had a powerful language structure, better readability, and excellent documentation, the implementation of SIMULA 67 for both compilation and execution was superior to SIMSCRIPT II at the time of their analysis.

A.4 APPLICATIONS OF SIMULATION MODELS

Simulation models are used for problems in many areas, including:

1. Engineering,

2. Military war game problems,

3. Operations research,

4. Economics,

5. Industrial production,

6. Biology and life sciences,

7. Sociology and political sciences,

8. Administrative and government planning.

It would be a formidable task to enumerate all or most of the simulation studies that have been conducted. The reader may get some idea of the vastness of their applications by perusing the journal *Simulation* published by the Society for Computer Simulation, headquartered in La Jolla, California. Our objective is to merely illustrate potential areas of applicability of the two radically different simulation techniques. The first, *system dynamics*, exemplifies a continuous deterministic modeling technique and has been extensively used in the study of population, environment, industrial production, natural resources, and other socio-economic problems. The second is a discrete stochastic modeling technique which has been used to evaluate service systems such as production systems, transport, and public administration.

A.4.1 System Dynamics

The system dynamics method has been developed at the Massachusetts Institute of Technology during more than 30 years of effort directed toward the analysis and control of complex system behavior. According to the theory underlying system dynamics, the method is applicable to systems in which only two types of variables are necessary to express internal relationships. They are: (1) levels and (2) rates.

Levels are the state variables, i.e., quantities that characterize the system at any point of time, such as population, natural resources, industrial output, labor demand, or capital stock. Levels are altered by rates of changes, such as birth rate, migration rate, natural resource consumption rate, etc. Rates reflect goals and decisions taken in the system, and are generated by the state of the system, which in turn is affected by the rates of change. This mutual relationship between rates and levels allows us to model systems that form feedback loops. A simplified model of animal population growth will illustrate the essential concepts of system dynamics. In describing the model, we follow the assumptions made and results obtained by Calhoun [1962]. Goodman [1974] used a system dynamics model to formalize this study and reproduced its results.

Calhoun investigated the effect of crowding on infant mortality and experimented with a rat colony confined to a limited area of land. No migration or predation was allowed. The environment was relatively constant to eliminate unwanted effects, and the rats had a sufficient supply of food. Calhoun observed that the colony size stabilized at a certain level. The population growth had terminated due to the very high infant mortality caused by overcrowdedness. Mother rats did not perform their maternal duties well in the overcrowded environment.

Figure A.4.1 shows a causal-loop diagram for the rat population model. This loop identifies the principal quantities and relationships between them and expresses our hypotheses about the functioning of the system. The variables are linked pairwise by arrows with plus or minus signs which indicate how a change in one variable influences other variables, assuming that all other factors are kept unchanged. For instance, an increase in the rat population increases the density, which in turn decreases births. This causal-loop diagram can now be translated to a more rigorous flowchart, shown in Fig. A.4.2, where the rectangular box represents a level and valve-like boxes represent rates. Auxiliary relationships between variables appear in the circles.

Using system dynamics nomenclature, we can describe the system, shown in Fig. A.4.2, by one level equation and two rate equations. The level difference equation is:

$$x_{t+1} = x_t + \Delta t(B_t - D_t), \qquad (A.4.1)$$

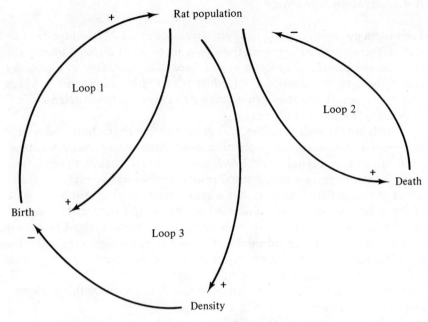

Figure A.4.1 Causal diagram.

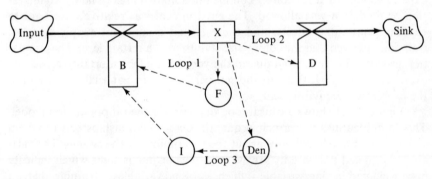

Figure A.4.2 Rat population flowchart.

where

x_{t+1} = Rat population at the time interval $t + 1$,
x_t = Rat population at the time interval t,
Δt = Time increment,
B_t = Birth rate, period t,
D_t = Death rate, period t.

The values of B_t and D_t are computed from the rate equations:

$$B_t = C_1 * F_t * I(DEN_t) \qquad (A.4.2)$$

$$D_t = \frac{x_t}{C_2} \qquad (A.4.3)$$

where

C_1 = Normal rate fertility, constant in time,

F_t = Female rat population,

$I(DEN_t)$ = Infant survival multiplier, experimentally established function of the population density DEN_t, shown in Fig. A.4.3,

C_2 = Average rate lifetime, constant.

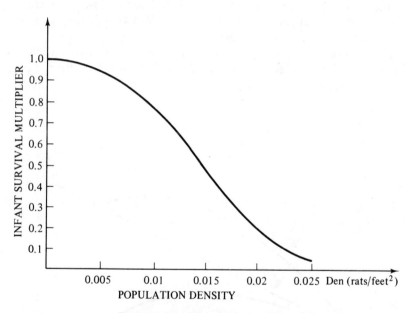

Figure A.4.3 Infant survival multiplier.

Furthermore, we assume that half of the population are females

$$F_t = 0.5 * x_t \qquad (A.4.4)$$

and that the population density is proportional to x_t,

$$DEN_t = C_3 * x_t \qquad (A.4.5)$$

where C_3 is the inverse of the area of confinement. Difference equation (A.4.1) and Eq . (A.4.2)–(A.4.5) completely specify our growth model. An alternative formulation can be given by the following differential equation:

$$\dot{X} = 0.5 * C_1 * X * I(C_3 * X) - \frac{X}{C_2} \qquad (A.4.6)$$

Assuming that $I(C_3 * X)$ is constant, $I = C_4$, equation (A.4.6) would be simply

$$\dot{X} = (0.5 * C_1 * C_4 - C_2^{-1})X \qquad (A.4.7)$$

whose solution is exponential and if

$$0.5 * C_1 * C_4 - C_2^{-1} > 0 \qquad (A.4.8)$$

then the rat population would increase without bound. However, increasing population density impacts the birth rate (Fig. A.4.4), and the net growth rate $B - D = 0$ for X_e which is the rat population size in steady state. The model can be further refined by introducing more realistic assumptions and by accounting for the effects of age on fertility, finite supply of food, disease, natural enemies, etc.

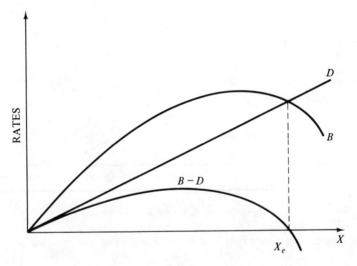

Figure A.4.4 Population growth and death rates.

These additions would increase the model size but would not change the nature of the modeling process that has been shown.

Differential equations developed by system dynamics can be solved by any standard differential equation solver. However, to help the user in his solution process, a special language called DYNAMO has been developed to accept the difference equations for models of dynamic feedback systems and to produce the desired numerical results in the form of tables and graphical plots. The DYNAMO language is discussed in Section A.3.1.

For more information on system dynamics, the reader is referred to Forrester [1961] and [1973], and Meadows, et al. [1973].

A.4.2 Evaluation of Service Systems

Many studied prototype systems in the area of production and management sciences involve processes in which service is provided on demand to randomly arriving entities such as customers, vehicles, etc. This demand for service has to be balanced against the limited resources available to satisfy the need. The usual objective of a study is to find the minimum amount of resources and operation rules that will satisfy an assumed demand. Alternatively, we may want to optimize the service given fixed resources. One of the disciplines of mathematics dealing with such problems is queuing theory which enables us to solve analytically some relatively simple cases. If it can be applied at all, queuing theory gives explicit solutions and enables one to find the optimal operation of the modeled service system.

Unfortunately, most queuing models reflecting properties of real-life systems are not amenable to mathematical analysis and cannot be solved in closed form. In these complex cases we have to apply computer simulation techniques and analyze numerically various queuing situations.

Service systems provide service to customers arriving at a given rate. Due to the stochastic nature of arrivals and service time, it may be necessary to form a queue of customers waiting for service. The queue is a result of the variability of arrivals and service in the system.

A stochastic service system can be characterized by the following three elements:

1. Statistical properties of the arrival pattern. The entities, such as customers, arrive randomly but we assume that the interarrival time follows a certain probability density function.

2. Statistical properties of the service process. It can be described in terms of service capacity, such as the number of servers, and the probability density function defining the distribution of service times.

3. The queue discipline which defines rules of queue formation and selection of entities for service.

Let us first consider a simple service system modeling problem that can be solved analytically.

Our model is based on the following assumptions:

1. There is only one queue and one server,

2. The queue discipline is first-come, first-served,

3. The arrival process is Poisson, i.e., the probability density function for the time between any two successive arrivals is

$$f(t) = \lambda e^{-\lambda t}, \quad t \geq 0 \tag{A.4.9}$$

where λ = Arrival rate per unit of time, and

4. The server processes one customer at a time. It is assumed that the probability density function for the length of time to serve one customer is

$$g(t) = \mu e^{-\mu t}, \quad t \geq 0 \tag{A.4.10}$$

where μ = Service rate per unit of time that the server is busy.

Assumption (3) above implies that the time of the next arrival is independent of the previous arrivals and the probability of arrival in a small interval Δt is proportional to $\lambda \Delta t$. For many real-life service systems this assumption is quite adequate.

Assume that the system starts its operation at $t = 0$.

The probability that no arrival occurs in $(0, T)$ is the same as the probability that the interarrival time is not less than T,

$$P[t \geq T] = \int_T^\infty \lambda e^{-\lambda t} \, dt = e^{-\lambda T} \tag{A.4.11}$$

The conditional probability that no arrival occurs in $(0, T + \Delta t)$ given that no arrival occurs in $(0, T)$ is

$$\frac{\int_{T+\Delta t}^\infty \lambda e^{-\lambda t} \, dt}{\int_T^\infty \lambda e^{-\lambda t} \, dt} = e^{-\lambda \Delta t} = P[t \geq \Delta t] \tag{A.4.12}$$

This probability depends only on Δt regardless of the time of last arrival. The arrival process can be, therefore, considered as memoryless.

The second property can be demonstrated in rough terms by considering the probability that an arrival occurs in an interval of length Δt.

$$P = 1 - e^{-\lambda \Delta t} \tag{A.4.13}$$

which can be approximated for small enough Δt by

$$P = 1 - (1 - \lambda \Delta t) = \lambda \Delta t \tag{A.4.14}$$

A similar interpretation can be developed for the exponential service pattern given by (A.4.10).

The following quantities are the principal measures of queues:

N_s = Mean number of entites in the system,
N_q = Mean number of entities in the queue (mean queue length),
T_s = Mean time of complete service (waiting and servicing),
T_q = Mean time spent waiting in the queue.

All these quantities can be established analytically for the considered queue. Let $P_n(t)$ be the probability that n entities are in the system at the time t. We assume again that $t = 0$ is the system's starting point and that Δt is a small time interval, such that the probability of more than one arrival or departure in Δt can be ignored.

If there are n entities in the system at time $t + \Delta t$ then we consider only the possibilities that there were either $n - 1, n$ or $n + 1$ entities in the system at time t. Therefore, for $n > 0$ the probability $P_n(t + \Delta t)$ can be approximated by

$$P_n(t + \Delta t) = P_{n-1}(t)(1 - \mu\Delta t)\lambda\,\Delta t + P_n(t)\mu\,\Delta t\,\lambda\,\Delta t \\ + P_n(t)(1 - \mu\,\Delta t)(1 - \lambda\,\Delta t) + P_{n+1}(t)(1 - \lambda\,\Delta t)\mu\,\Delta t \tag{A.4.15}$$

The terms on the righthand side of (A.4.15) correspond to the following possibilities:

1. $n - 1$ entities at time t, one arrival, no departure in Δt,
2. n entities at time t, no arrival, no departure in Δt,
3. n entities at time t, one arrival, one departure in Δt,
4. $n + 1$ entities at time t, no arrival, one departure.

Taking $P_n(t)$ to the lefthand side, dividing the equation by t and letting $\Delta t \to 0$, we get the differential equations

$$\dot{P}_n(t) = \lambda P_{n-1}(t) - (\lambda + \mu)P_n(t) + \mu P_{n+1}(t) \quad \text{for } n = 1, 2, 3, \ldots \tag{A.4.16}$$

and

$$\dot{P}_0(t) = -\lambda P_0(t) + \mu P_1(t), \quad n = 0 \tag{A.4.17}$$

Equations (A.4.16) and (A.4.17) represent the probabilities during the transient period of the queuing process. The solutions to these equations are functions of time and the probabilities of the number of customers in the system at time $t = 0$.

To study the queue in the steady state we set $\dot{P}_n(t) = 0, n = 0, 1, 2, \ldots$ and obtain the difference equations

$$\lambda P_{n-1} - (\lambda + \mu)P_n + \mu P_{n+1} = 0, \quad n = 1, 2, \ldots \tag{A.4.18}$$

$$-\lambda P_0 + \mu P_1 = 0, \quad n = 0 \tag{A.4.19}$$

Equations (A.4.18) and (A.4.19) are homogeneous and their solution is

$$P_n = \rho^n P_0, \quad n = 0, 1, 2, \ldots \tag{A.4.20}$$

where

$$\rho = \frac{\lambda}{\mu} \tag{A.4.21}$$

We now assume that

$$\rho < 1 \tag{A.4.22}$$

i.e., that the service rate exceeds the rate of arrival. Without this assumption, the queue length would grow without a bound.

From

$$\sum_{n=0}^{\infty} P_n = P_0 \sum_{n=0}^{\infty} \rho^n = \frac{P_0}{1 - \rho} = 1 \tag{A.4.23}$$

we get

$$P_0 = 1 - \rho \tag{A.4.24}$$

and

$$P_n = (1 - \rho)\rho^n \tag{A.4.25}$$

The mean number of entities in the system can now be computed as

$$N_s = \sum_{n=0}^{\infty} n P_n = (1 - \rho)\rho \sum_{n=1}^{\infty} n \rho^{n-1} \tag{A.4.26}$$

since

$$\sum_{n=1}^{\infty} n \rho^{n-1} = \sum_{n=1}^{\infty} \frac{d}{d\rho}(\rho^n) = \frac{d}{d\rho} \sum_{n=1}^{\infty} \rho^n = \frac{d}{d\rho}\left(\frac{\rho}{1 - \rho}\right) = \frac{1}{(1 - \rho)^2} \tag{A.4.27}$$

and

$$N_s = \frac{\rho}{1 - \rho} \tag{A.4.28}$$

Since the queue length differs from the number of entities in the system by one entity, we have

$$N_q = 0 \cdot P_0 + \sum_{n=1}^{\infty} (n - 1)P_n = \sum_{n=1}^{\infty} n P_n - \sum_{n=1}^{\infty} P_n$$

$$N_q = \frac{\rho}{1 - \rho} - (1 - P_0) = \frac{\rho}{1 - \rho} - \rho = \frac{\rho^2}{1 - \rho} \tag{A.4.29}$$

If an entity arrives when there are n entities in the system, its waiting time before service begins is the time of service of those n entities. This waiting time is n/μ since $1/\mu$ is the average service time for one entity. The mean time ^0T_q is computed from

$$T_q = \sum_{n=0}^{\infty} \frac{n}{\mu} P_n = \frac{1}{\mu} \sum_{n=0}^{\infty} nP_n = \frac{N_s}{\mu} = \frac{1}{\mu} \frac{\rho}{1-\rho} \qquad (A.4.30)$$

The mean time T_s is larger by $1/\mu$,

$$T_s = T_q + \frac{1}{\mu} = \frac{1}{\mu - \lambda} \qquad (A.4.31)$$

Equations (A.4.30)–(A.4.31) are simple and useful measures of performance for our service system.

In more realistic service systems we may have some of the following additional complications:

1. Several servers arranged in parallel or in series; waiting customers form either one queue or several queues.

2. If there are several queues, customers may switch queues before they proceed to a server.

3. A customer seeing a long queue may balk and leave the system.

4. Customers may also change their minds and decide to leave after joining the queue; the maximum queue length may be limited by available space.

Thus the arrival may depend on the status of the queuing system. Another important factor that should be taken into account in more accurate studies is the variability of the service rate with changing queue length. Figure A.4.5 shows a model incorporating features (3)–(5) of the service system. Some of these modifications would make the queuing system nontractable analytically. It is particularly difficult to get analytical expressions for the transient period of service. It would also be very difficult, if not impossible, to obtain analytical results considering service systems with the capability to switch abruptly to a different set of parameters, e.g., activate a new server if the queue exceeds a certain length.

The stochastic service system with such complexities can be modeled and investigated numerically by using simulation techniques. In such cases we construct a discrete simulation model using one of the available discrete simulation languages and run numerical experiments to obtain quantitative information related to:

1. A measure of provided service,

2. The amount of resources that have been used to meet the demand.

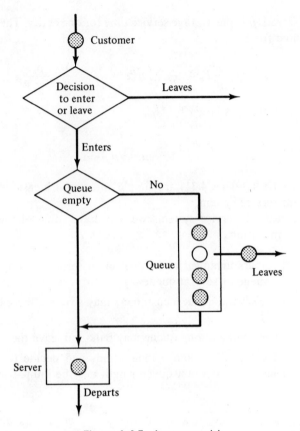

Figure A.4.5 A queue model.

The provided service can be evaluated on the basis of a frequency histogram of the waiting times for customers. Several measures can be used to evaluate resources used to meet the demanded service. They include:

1. A time distribution histogram for the number of occupied servicing units,

2. A time distribution histogram for the size of the queue,

3. A frequency histogram for the idle times of servicing units.

The chief advantage of numerical simulation of service systems is the ease of modifying assumptions, introducing nonstandard distribution functions or any complex conditions under which the system has to perform. On the other hand, the numerical simulation does not provide the modeler with the optimal operation or size of the system. It usually takes many runs to cover the ranges of interest of input variables and identify near-optimal solutions.

The reader interested in practical applications of queuing processes is referred to Saaty [1961]. A more recent book by Wagner [1970] contains a chapter on queuing theory written from a modeling point of view.

A.5 SUMMARY

We have described continuous and discrete simulation models and several computer simulation languages. Tables A.5.1 and A.5.2 provide a list of emulators and expression-based languages for initial value problems, and languages for partial differential equations.

Table A.5.1 Emulators and Expression-based Simulation Languages for Initial Value Problems

LANGUAGE TYPE	COMPUTER LANGUAGES
Emulators	MIDAS, MADBLOCK, CSMP 1130
Expression-based	MIMIC, DARE P, CSMP/360, CSMP-III, CSSL-III, PROSE, DYNAMO, ACSL, EASY

Table A.5.2 Languages for Partial Differential Equations

COMPUTER LANGUAGE	EQUATION TYPE
PDEL	Elliptic, parabolic, hyperbolic
LEANS	Elliptic, parabolic, hyperbolic
DSS	Elliptic, parabolic
PDLAN	Parabolic (meteorology)
FORSIM	Elliptic, parabolic, hyperbolic

Table A.5.3 lists some well-known discrete simulation languages and the main features of prototypes for which they can be used. A more detailed decision flow diagram which may serve as an aid in selecting a suitable simulation language is given in Shannon [1975]. We may add, however, that there is considerable overlap in the capabilities of these languages, and very often selection of a language is based on such factors as user familiarity with it and the availability of a particular compiler.

Table A.5.4 summarizes our comparison between the leading three languages for discrete simulation: GPSS, SIMULA, and SIMSCRIPT.

Table A.5.3 Discrete-system Simulation Languages

LANGUAGE TYPE	PROTOTYPE MAIN FEATURES	COMPUTER LANGUAGE
Event oriented	Large number of entities, moderately interacting processes, fixed time incrementing mechanism.	SIMSCRIPT II, GASP
Activity oriented	Fixed number of entities, highly interacting processes, events taking place irregularly.	CSL, MILITRAN
Process oriented	Large number of interacting processes, prototype is defined by many chains of actions together with the ways in which each process interacts with others.	SIMULA 67, ASPOL, SOL
Transaction-flow oriented	Initial study of inventory and queuing problems.	GPSS

Table A.5.4 GPSS, SIMULA, SIMSCRIPT comparison

FEATURE	GPSS	SIMULA	SIMSCRIPT
Versatility	Restricted	General	General
Computational facilities	Poor	Good	Good
General-purpose programming facilities	No	Yes	Yes
Ease of use	Easy	For advanced users	For advanced users
Computational efficiency	Low	High	High
Data collection facilities	Adequate	Good	Excellent
Input-Output facilities	Inflexible	Good	Good
Compiler availability	Good	Restricted (in USA)	Very good

A.6 SOURCES OF INFORMATION RELATED TO SIMULATION LANGUAGES

ACSL:

ACSL User's Guide/Reference Manual, Mitchell and Gauthier, Associates, 1337 Old Marlboro Road, Concord, Mass. 01742.

Bover, D. C. C., and M. R. Osborne, *Introduction to Continuous Simulation Using ACSL*, Computer Centre, Australian National University, Tech. Rep. No. 58, 1977.

CSL:

Control and Simulation Language Reference Manual, IBM United Kingdom, Ltd., and Esso Petroleum Co., Ltd., March, 1963.

CSMP-III:

Continuous System Modeling Program III and Graphic Feature, General Information Manual, a program product number 5734-X59, Manual Number GH 19-7000, IBM, White Plains, N.Y., 1971.

CSMP 1130:

1130 CSMP Program Reference Manual, IBM 1130-CX-13X.

CSMP SYSTEM 360:
  Continuous System Modeling Program, User's Manual, a type II program
  number 360A-CX-16X, Manual Number GH 20-0367, IBM, White
  Plains, N.Y., 1968.

CSSL-III:
  User's Guide and Reference Manual, Programming Sciences Corporation,
  Los Angeles, CA, 1970.

  Continuous System Simulation Language Version 3, User's Guide, Control
  Data Corporation, Sunnyvale, CA, 1971.

DARE P:
  Lucas, J. J., and J. V. Wait, *DARE P User's Manual*, CSRL Report 255,
  University of Arizona, College of Engineering, Computer Science
  Research Laboratory.

DSS:
  Zellner, M. G., *DSS-Distributed System Simulator*, Ph.D., Lehigh
  University, Bethlehem, PA, 1970.

DYNAMO:
  Pugh, A. L., *DYNAMO II User's Manual*, MIT Press, Cambridge,
  Mass., 1970.

  Forrester, J. W., *Principles of Systems*, Wright-Allen Press, Inc., Cambridge, Mass. 1973.

EASY:
  *EASY, Environmental Control System—Transient Analysis, Vol. III:
  Integrated Transient Computer Program Users' Guide*, Flight Dynamics
  Laboratory: AFFDL-TR-77-102, Wright-Patterson AFB, Ohio, 1977.

  Application of the EASY Dynamic Analysis Program to Aircraft Modeling,
  Boeing Computer Services, Seattle, Wash., Document No. 40120, 1976.

ECSS:
  Nielsen, N. R., *An Extendible Computer System Simulator*, The Rand
  Corporation, RM-6132-NASA, 1970.

FORSIM:
  Carver, M. B., A FORTRAN-Oriented Simulation System for the
  General Solution of Partial Differential Equations, *Proceedings of
  the 1973 Summer Computer Simulation Conference*, Montreal, July,
  1973.

GASP:

Pritsker, A. A. B., and P. J. Kiviat, *Simulation with GASP II: A FORT-RAN Based Simulation Language*, Prentice-Hall, Inc., Englewood Cliffs, N.J., 1969.

GPSS:

Gordon, G., *The Application of GPSS V to Discrete System Simulation*, Prentice-Hall, Inc., Englewood Cliffs, N.J., 1975. *General Purpose Simulation System/360 User's Manual*, IBM H20-0326, 1967.

LEANS:

Schiesser, W. E., *An Introduction to LEANS-III Lehigh Analog Simulator Version III and DSS Distributed System Simulator*, Lehigh University, Bethlehem, PA, Computer Center Report, 1973.

MADBLOCK:

Rideout, V. C., and L. Tavernini, "MADBLOCK, A Program for Digital Simulation of a Hybrid Computer," *Simulation*, 20, January, 1965.

MIDAS:

Blechman, G. E., "An Enlarged Version of MIDAS," *Simulation*, 41, October, 1964.

Burgin, G. H., "MIDAS III, A Compiler Version of MIDAS," *Simulation*, 160, March, 1966.

MILITRAN:

MILITRAN Reference Programming and Operations Manuals, Systems Research Group, Inc.

MIMIC:

Peterson, H. E., and F. J. Sansom, *MIMIC Programming Manual*, Air Force Systems Command, Wright-Patterson Air Force Base, Ohio, 1967.

MIMIC Digital Simulation Language, Reference Manual, Control Data Corporation, Sunnyvale, CA, 1968.

PDEL:

Cardenas, A. F., *The PDEL Language and Its Use to Solve Partial Differential Equation Problems*, University of California, Computer Science Department Report, Los Angeles, CA, June, 1972.

PDLAN:
   Helgason, R., and J. Gary, *PDLAN User's Manual Version 1*, National
   Center for Atmospheric Research, Boulder, CO, 1971.

PROSE:
   Thames, J. M., *PROSE—A Problem Level Programming System*, Solve-
   ware Associates, San Pedro, CA, 1973.

QUICKSIM:
   Weamer, D. G., "QUICKSIM—A Block Structured Simulation Lan-
   guage Written in SIMSCRIPT," *Third Conference on Application of
   Simulation*, 1, December, 1966.

SIMSCRIPT:
   Kiviat, P. J., R. Villanueva, and H. M. Markowitz, *SIMSCRIPT II.5
   Programming Language*, Consolidated Analysis Centers, Inc., Los
   Angeles, CA, 1973.

   SIMSCRIPT II.5, Reference Manual, Consolidated Analysis Centers,
   Inc., 1972.

   SIMSCRIPT II.5 User's Manual S/360-370 Version, Consolidated
   Analysis Centers, Inc., 1974.

SIMULA:
   Birtwistle, G. M., O. J. Dahl, B. Myrhaug, and K. Nygaard, *SIMULA
   BEGIN*, Auerbach Publishers, Philadelphia, PA, 1973.

SOL:
   Knuth, D. E., and J. L. McNeley, SOL—A Symbolic Language for
   General Purpose System Simulation, *IEEE Transactions on Electronic
   Computers*, 401, 1964.

REFERENCES

BEKEY, G. A. and W. J. KARPLUS, *Hybrid Computation*. New York: Wiley, 1968.

BENYON, P. R., "A Review of Numerical Methods for Digital Simulation," *Simulation*, 219, 1968.

CALHOUN, B. F., "Population Density and Social Pathology," *Scientific American*,
206, 1962.

EMSHOFF, J. R., and R. L. SISSON, *Design and Use of Computer Simulation Models*.
London: MacMillan Co., Collier-MacMillan, Ltd., 1970.

FORRESTER, J. W., *Principles of Systems*. Cambridge, Mass.: Wright-Allen Press, 1973.

FORRESTER, J. W., *Industrial Dynamics*. Cambridge, Mass.: M.I.T. Press, 1961.

GEAR, C. W., *Numerical Initial Value Problems in Ordinary Differential Equations*. Englewood Cliffs, N.J.: Prentice-Hall, 1971.

GOODMAN, M. R., *Study Notes in System Dynamics*. Cambridge, Mass.: Wright-Allen Press, 1974.

GORDON, G. *The Application of GPSS V to Discrete System Simulation*. Englewood Cliffs, N.J.: Prentice-Hall, 1975.

HILLS, R., *SIMULA 67, An Introduction*, Norwegian Computing Center, Oslo, Norway, 1972.

HILLS, R., *An Introduction to Simulation Using SIMULA*, Norwegian Computing Center, Oslo, Norway, 1973.

HINDMARSH, A. C., "GEAR: Ordinary Differential Equation Solver," *Rep. UCID-30001*, Rev. 2, Lawrence Livermore Lab., Livermore, CA, 1972.

HINDMARSH, A. C., "GEARB: Solution of Ordinary Differential Equations Having Banded Jacobian," *Rep. UCID-30059*, Lawrence Livermore Lab., Livermore, CA, 1973.

KARPLUS, W. J. and R. N. NILSEN, "Continuous System Simulation Languages: A State-of-the-Art Survey," *Annales de l'Association Internationale pour le Calcul Analogique*, no. 1, January 17, 1974.

KLEINE, H., "A Second Survey of User's Views of Discrete Simulation Languages," *Simulation*, vol. 17, no. 2, 1971.

KROGH, F. T., "Algorithms for Changing the Step Size," *SIAM J. Numer. Anal.*, 10, 5, 1973.

MEADOWS, D. L., and D. H. MEADOWS, (eds.), *Toward Global Equilibrium: Collected Papers*, Cambridge, Mass.: Wright Allen Press, 1973.

ORD-SMITH, R. J. and J. STEPHENSON, *Computer Simulation of Continuous Systems*, New York: Cambridge University Press, 1975.

PALME, J., *Simula 67—An Advanced Programming and Simulation Language*, Norwegian Computing Center, Oslo, Norway, 1970.

PRITSKER, A. A. B., and P. J. KIVIAT, *Simulation with GASP II*. Englewood Cliffs, N.J.: Prentice-Hall, 1969.

ROGEBERG, T., *Simulation and Simulation Languages*, Norwegian Computing Center, Oslo, Norway, 1973.

SAATY, T. L., *Elements of Queuing Theory, with Applications*, New York: McGraw-Hill, 1961.

SHANNON, R. E., *System Simulation, The Art and Science*, Englewood Cliffs, N.J.: Prentice-Hall, 1975.

SINCOVEC, R. F. and N. K. MADSEN, "Software for Nonlinear Partial Differential Equations", *ACM Transactions on Mathematical Software*, vol. 1, no. 3, 1975.

TOGNETTI, K. P., "Discrete Simulation Languages with Reference to Bio-simulation —A User's Impressions," *Proceedings of Fourth Australian Computer Conference*, Adelaide, 1969.

TOGNETTI, K. P., and C. BRETT, SIMSCRIPT II and SIMULA 67—A Comparison," *The Australian Computer Journal*, vol. 4, no. 2, 1972.

VIRJO, A., *A Comparative Study of Some Discrete Event Simulation Languages*, Finnish State Computing Centre, Helsinki, Finland, 1972.

WAGNER, H. M., *Principles of Management Science*, Englewood Cliffs, N.J.: Prentice-Hall, 1970.

ZEIGLER, B. P., *Theory of Modelling and Simulation*, New York: John Wiley, 1976.

Index

A

ACSL, 255, 279, 281
Adaptation (*see* Computer program,
    adaptation)
Algorithm, 63-71
    convergence, 66, 67, 87, 129
    convergence order, 29, 67
    direction, 112, 123, 127, 223
    efficiency, 29
    robust performance, 67, 73
    space complexity, 67
    stability, 68, 73
    step, 65-66, 112, 128, 223
    termination, 65
    time complexity, 67
Approximation, finite difference (*see* Finite-
    difference approximation)
Approximation, initial (*see* Initial approx-
    imation)
Approximation error (*see* Error, approxima-
    tion)
Approximations (in mathematical model),
    26, 33, 69, 238
ASPOL, 280
Asymptotic error constant (*see* Error,
    asymptotic constant)

B

Balance equation, 48, 52
Basic requirements (*see* Computer program,
    requirements)
Bottom-up design (*see* Computer program,
    design, bottom-up)
Boundary conditions (*see* Conditions,
    boundary)
Boundary value problem, 65, 82, 96 (*see
    also* Conditions, side)

C

Cancellation error (*see* Error, cancellation)
CDC Math Science Library, 72
Choleski method, 88
Circuit, 16, 54
Circuit law (*see* Kirchhoff laws)
Code (*see* Computer code)
Coding (*see* Computer program, coding)
Comment (*see* Computer program, comment)
Complexity (*see* Algorithm, space complexity,
    time complexity)
Complex model (*see* Mathematical model,
    complex)
Component interaction matrix, 46, 48, 54,
    55
Computer code, 163, 191, 201-7, 213
Computer implementation (*see* Computer
    program, and Mathematical model,
    building process, computer imple-
    mentation)
Computer program:
    adaptation, 144, 184-87
    bottom level, 149
    change, 183, 213
    coding, 140, 199, 201-7
    comment, 183, 202, 204, 205
    configuration control, 146, 182, 209-11,
        214
    construction, 144, 199-201
    control:
        interface, 191
        logic, 155
        statement, 161, 171-72, 176
    design, 143-44, 146-99
        adaptation (*see* Computer program,
        adaptation)
        bottom-up, 201

Computer program *(cont.):*
  design *(cont.):*
    correctness *(see* Computer program, proof of correctness)
    criteria, 148
    detailed, 143
    documentation, 179–83
    error, 147, 163–64 *(see also* Error, computer program)
    evaluation, 174–79
    level, 149, 160, 164, 168, *(see also* Computer program, bottom level, and top level
    methodology, 146
    model, 148–49, 163–64, 180
    module, 177–78, 181, 189, 199, 201, 208, 210
    module name, 152–53, 181–83
    path, 160, 168–73, 176, 207–8
    path selection, 160–63
    preliminary, 143
    representation, 150
    requirements *(see* Computer program, requirements)
    review, 183–84
    steps, 150–51
    testing, 163–74
    top-down (TDD), 146–51, 180
    tree, 152, 159, 176–78, 183, 188
  execution time, 175–76, 187
  interfaces, 144, 190–91
  maintenance, 145–46, 156, 211–14
  maintenance document, 199
  minimum configuration, 144, 148, 199–201
  operational sequence, 151, 154, 157–63, 165
  proof of correctness, 149, 164–74
  requirements, basic, 142, 180, 193
  requirements, supplemental, 143, 193
  requirements tree, 149
  segmentation, 188–90
  standard format, 202
  storage, 177–78, 187, 190
  termination, 173, 207–8
  testing, 145, 207–8
  top level, 149
  useability, 178
Computing environment, 144, 185
Condition *(see* Mathematical modeling problem, condition)
Conditions:
  boundary, 19, 26, 55, 82, 96, 194
  initial, 19, 26, 55, 116
  necessary and sufficient for model usefulness, 9, 21, 27, 41, 44
  necessary for computer program design correctness, 165, 172–74
  side, 19, 40, 55 *(see also* Boundary value problem)
  terminal, 19, 40, 54, 55
Condition statements, 162

Conductivity, 15, 50, 79, 80
Configuration control *(see* Computer program, configuration control)
Conjugate gradient method, 124, 126
Connection matrix, 165, 167–71
Connectivity analysis, 165, 171
Conservation of flow, 97, 100, 117
Conservation principle, 48–49, 52, 55
Constraints, 119, 149, 158
Continuous simulation *(see* Simulation, continuous)
Continuous simulation languages *(see* Simulation, languages, continuous)
Control:
  area, 81
  elements, 17
  space, 52
  system models, 17, 48
  variable *(see* Variables, control)
  volume, 47
Controlled elements, 17
Convergence *(see* Algorithm, convergence)
Convergence error *(see* Error, convergence)
Criterion of merit *(see* Optimization objective)
CSL, 280, 281
CSMP 1130, 253, 279, 281
CSMP SYSTEM 360, 254, 279, 282
CSMP-III, 254, 279, 281
CSSL-III, 254, 279, 282

D

Darcy's law of water seepage, 81
DARE P, 253, 279, 282
Data, 12, 26–27, 69, 144, 184–87, 190–91, 233, 238, 250, 260
Data error, *(see* Error, data)
Decomposition *(see* Process decomposition)
Design *(see* Computer program, design)
Design, detailed *(see* Computer program, design, detailed)
Design, top down *(see* Computer program, design, top-down)
Detailed design *(see* Computer program, design, detailed)
Directed graph, 157, 160
Directed line, 172
Direction, *(see* Algorithm, direction)
Dirichlet problem, 82, 96
Discrete simulation *(see* Simulation, discrete)
Discrete simulation languages *(see* Simulation languages, discrete)
Dissipativity, 15, 50, 79
Dissipator, 15, 50, 56
Distributed parameter model *(see* Mathematical model, distributed parameter)
DSS, 257, 279, 282
DYNAMO, 255, 272, 279, 282

E

EASY, 256, 279, 282
ECSS, 261, 282
Edge, 160-62
EISPACK, 75-76
Elements:
  control (*see* Controlled elements)
  controlled (*see* Controlled elements)
  feedback (*see* Feedback elements)
  model (*see* Mathematical model, elements)
Energy, 47, 50, 118
Entry, 158
Entry point, 158, 204
Error:
  analysis, backward, 230
  approximation, 86
  asymptotic constant, 67
  cancellation, 70, 229-30
  computational (see Error, round-off)
  computer program, 148, 154, 163, 169,
    171, 200, 207-8, 211-13 (*see also*
    Computer program, design, error)
  convergence, 230-31
  data, 26, 61
  discretization, 231-32
  estimate, 70
  mathematical, 229
  message, 154
  round-off, 70, 86, 229-30
  truncation, 69
Evaluation (*see* Computer program, design,
  evaluation)
Execution time (*see* Computer program,
  execution time)
Exit, 158, 204
Expressions, logical (*see* Logical expressions)
Expressions, mathematical (*see* Mathematical
  expressions)

F

Feedback elements, 17
Finite-difference approximation, 86
Finite elements, 90
Fixed parameters (*see* Parameters, fixed)
Fletcher-Reeves method, 124
Flow chart, 160
FORSIM, 258, 279, 282
Fourier's law of heat conduction, 81
Functional correspondence (*see* Input/out-
  put functional correspondence)

G

GASP, 280, 283
Gauss elimination method, 88
Gauss-Seidel method, 88
GO TO statement, 171
GPSS, 263-64, 267, 280, 281, 283

Graph, directed (*see* Directed graph)
Graph connectivity analysis (*see* Connectivity
  analysis)
Graph model (*see* Computer program,
  operational sequence)

H

Harwell Subroutine Library, 74-75
Hazen-Williams formula, 98
Hydraulic head, 96

I

Ill condition (*see* Mathematical modeling
  problem, condition)
Implicit loop procedure, 105-11
IMSL, 72
Induced inaccuracy, 226
Initial approximation, 65
Initial conditions (*see* Conditions, initial)
Input, 148, 150, 155-56, 160-63, 165, 172-
  73, 180, 190-91
Input/output functional correspondence,
  155-56, 180, 191
Input/output functional correspondence
  matrix, 156
Interaction matrix, component (*see*
  Component interaction matrix)
Interfaces (*see* Computer program,
  interfaces)
Iterative process, 65 (*see also* Algorithm)
Iterative step (*see* Algorithm, step)

J

Jacobian matrix, 112

K

Kirchhoff laws, 15, 16, 19, 54-56, 97, 99

L

Laplace's difference equation, 86
Laplace's equation, 82
LEANS, 257, 279, 283
Level (*see* Computer program, design, level)
LINPACK, 76-77
List processing, 251
Logical expressions, 46
Logical statements, 12, 26
Loop, 16, 54, 173, 176
Loop law (*see* Kirchhoff laws)
Lumped parameter model (*see* Mathematical
  model, lumped parameter)

M

MADBLOCK, 253, 279, 283
Maintenance (*see* Computer program,
  maintenance)

Management, mathematical model building and modeling, 21-22
Mathematical error (*see* Error, mathematical)
Mathematical expressions, 12, 26, 46
Mathematical model, 7, 26
  adaptive parameter, 14
  analysis, 26-27, 42-57
  building process, 21-33
    analytical stage, 35-136
    computer implementation, 9, 30-31, 137-214
    management (*see* Management, mathematical model building and modeling)
    model validation (*see* Mathematical model, validation)
  causal, 12
  complex, 14, 16, 44-45
  continuous, 13
  decomposition, 233
  definition, 26-27, 42-57
  detail degree, 42, 55
  deterministic, 13
  discontinuous, 13
  discrete, 13
  distributed parameter, 14, 50-52, 79-80
  dynamic, 12
  elements, 11-12, 26, 45
  feedback, 17
  feedforward, 17
  fixed parameter, 14
  idealization, 69
  implementation, 237-39
  instantaneous, 12
  lumped parameter, 14, 50-52, 96, 116
  lumping, 233
  memory, 12
  multistage, 16
  noncausal, 12
  nonoptimizing, 20
  optimizing, 20, 115, 223, 235
  procedural, 244
  process, 13, 16-17
  quasi-steady-state, 12
  recursive, 13
  refinement, 12
  static, 12
  stationary, 12
  steady-state, 12
  stochastic, 13
  structure, 12-15, 55, 95, 116
    branching, 14
    cyclic, 15
    network, 15, 95
    tree, 14, 116
  time invariant, 12
  types, 12-21
  unsteady, 12
  uses, 8, 239

Mathematical model *(cont.):*
  validation, 31-33, 215-37
    empiricism, 218
    face validity, 219
    rationalism, 218
Mathematical modeling problem:
  analysis, 28-30, 57-63, 83-85, 99-105, 121-27
  condition, 9, 28, 36, 61, 68, 225
  definition, 140-42
  reformulation, 28-30, 62-63, 121-22
  sensitivity, 71
  solution development, 28-30, 63-71, 85-89, 105-13, 127-30
  solvability, 9, 58-61, 84-85, 102-5, 129
  statement, 25, 38-42
Mathematical software, 71-78, 188
Matrix:
  component interaction (*see* Component interaction matrix)
  connection (*see* Connection matrix)
  functional correspondence (*see* Input/output functional correspondence matrix)
  Jacobian (*see* Jacobian matrix)
  reachability (*see* Reachability matrix)
  stiffness (*see* Stiffness matrix)
Maximum/minimum principle, 84
MIDAS, 253, 279, 283
MILITRAN, 280, 283
MIMIC, 253, 279, 283
Minimum configuration (*see* Computer program, minimum configuration)
MINPACK, 77-78
Model, 1
  abstract, 142, 192
  analog computer, 6
  computer program (*see* Computer program, design, model)
  computational, 106, 143, 144, 194
  formal, 4
  iconic, 5
  material, 4
  mathematical, 7, 26 (*see also* Mathematical model)
  physical, 4
  preliminary definition, 56
  procedural (*see* Mathematical model, procedural)
  purposes, 3-11
  refinement (*see* Mathematical model, refinement)
  scale, 6
  simulation, 18
  space, 9, 33, 216, 238
  symbolic, 4, 6, 8
  types, 3-11
  usefulness, 9-11

Modeling project plan, 22, 42
Modeling purpose, 21, 40, 42

N

Necessary and sufficient conditions (*see* Conditions, necessary and sufficient for model usefulness)
Neuman problem, 82, 96
Newton method, 111, 123, 126
Newton-Raphson method, 111
Newton search direction, 123
Node law (*see* Kirchoff laws)
Node points, 90
Numerical problem types, 18

O

Objective function (*see* Optimization objective)
Operational sequence (*see* Computer program, operational sequence)
Optimality criterion (*see* Optimization objective)
Optimization model, 115
Optimization objective, 20, 119, 121, 235
Optimization problem, 120, 126
Order of convergence (*see* Algorithm, convergence order)
Output, 148, 150, 154–56, 160–63, 165, 172–73, 180–83, 191
Output/input dependency, 160, 173

P

Parameters, fixed, 11, 26, 46
Parent process, 149, 152, 154–55, 163
Path, 16, 54 (*see also* Computer program, design, path)
PDEL, 257, 279, 283
PDLAN, 258, 279, 284
Plan, for modeling project (*see* Modeling project plan)
Potential function, 79, 80, 83
Potential gradient, 80
Power, 46, 50–52
Precision, 229–30
Preliminary design (*see* Computer program, design, preliminary)
Pressure, 96
Problem (*see* Mathematical modeling problem)
Process decomposition, 151–56, 182
Program (*see* Computer program)
Proof of correctness (*see* Computer program, proof of correctness)
PROSE, 254, 279, 284
Prototype, 1, 40, 79, 94, 115–116
 behavior, 8
 features, 26

Prototype *(cont.)*:
 identification, 25, 38–42
 space, 9, 33, 40, 142, 216, 238, 244
Pseudocode, 160

Q

Quasi-Newton method, 112, 124, 126
Queuing, 273
QUICKSIM, 261, 284

R

Reachability matrix, 168–72
Reachability vector, 168
Recursive relationships, 221
Reference value (condition), 17
Reformulation, problem (*see* Mathematical modeling problem, reformulation)
Relaxation:
 factor, 88
 successive over, 88
Requirements (*see* Computer program, requirements)
Reservoir:
 flux, 15, 51, 56
 potential, 16, 52, 56
Residual, 105, 108, 112
Resistivity, 79, 80
Robust performance of algorithm (*see* Algorithm, robust performance)
Round-off error (*see* Error, round-off)

S

Segmentation, program (*see* Computer program, segmentation)
Sensitivity analysis, 234
Sequential machine, 160–61
Side conditions (*see* Conditions, side)
Simplification of models, 29, 233
Simpson's formula, 227
SIMSCRIPT, 260–62, 267, 280, 281, 284
SIMULA, 262–63, 267, 280, 281, 284
Simulation:
 activity, 245, 249
 activity approach, 250
 analog, 6, 244, 247, 248
 attribute, 245, 261
 continuous, 245–48
 digital, 244, 248
 discrete, 245, 248–51
 discrete, components, 249
 entity, 245, 249, 261
 event, 249, 261
 event scheduling approach, 250
 languages, bibliography, 281–84
 languages, continuous, 6, 253–58
  emulators, 253, 279

Simulation *(cont.)*:
   languages, continuous *(cont.)*:
      expression based, 253–56, 279
      partial differential equations, 256–58, 279
   languages, discrete, 259–68
      activity oriented, 250, 259
      event oriented, 250, 259
      process flow oriented, 250, 259–60
      transaction flow oriented, 260
   languages, summary, 279–81
   numerical, 244
   process interaction, 250
   time advancing, 249
Software libraries *(see* Mathematical software)
SOL, 280, 284
Solution *(see also* Mathematical modeling problem, solvability):
   algorithm *(see* Algorithm)
   existence, 9, 28, 36, 58, 84, 103
   finite difference, 85
   numerical approximation, 9, 29
   stability, 9, 28, 36, 58, 61, 84
   technique *(see* Algorithm)
   uniqueness, 9, 28, 36, 58, 60, 84, 99, 101, 129
Stability of algorithm *(see* Algorithm, stability)
State diagram, 160
Step *(see* Algorithm, step)
Stiffness matrix, 91
Storage *(see* Computer program, storage)
Subprocess, 151
   input, 155
   operational sequences, 157–63
   output, 154–55
Successive over relaxation *(see* Relaxation, successive over)
Supplemental requirements *(see* Computer program, requirements)

System dynamics, 268–69

T

Terminal conditions *(see* Conditions, terminal)
Termination *(see* Algorithm, termination, and Computer program, termination)
Testing, computer program *(see* Computer program, design, testing)
Time of program execution *(see* Computer program, execution time)
Top-down design (TDD) *(see* Computer program, design, top-down)
Truncation error *(see* Error, truncation)

U

Uniqueness *(see* Solution, uniqueness)
Useability *(see* Computer program, useability)

V

Validity, model *(see* Mathematical model, validation)
Variables, 11, 26, 46, 100
   across, 15, 46, 54
   control, 160, 173
   controllable, 16
   independent, position, 15, 80
   independent, spatial, 15
   independent, time, 15
   intermediate, 16
   performance, 16
   process model, 16
   switch, 155, 160
   through, 15, 46, 54
   uncontrollable, 16
   unknown, 16